The World of Small States

Volume 1

Series editors
Petra Butler
Wellington, New Zealand

Caroline Morris
London, United Kingdom

More information about this series at http://www.springer.com/series/15142

Petra Butler • Caroline Morris

Editors

Small States in a Legal World

 Springer

Editors
Petra Butler
Faculty of Law
Victoria University of Wellington
Wellington, New Zealand

Caroline Morris
School of Law
Queen Mary University of London
London, United Kingdom

The World of Small States
ISBN 978-3-319-39365-0   ISBN 978-3-319-39366-7   (eBook)
DOI 10.1007/978-3-319-39366-7

Library of Congress Control Number: 2017937531

Printed on acid-free paper

This Springer imprint is published by Springer Nature
The registered company is Springer International Publishing AG
The registered company address is: Gewerbestrasse 11, 6330 Cham, Switzerland

Preface

The World Bank Group defines small states as 'as countries that . . .have a population of 1.5 million or less. . .'.[1] Of the world's 195 commonly recognised sovereign states, 40 countries, or 21% of the total number of states, come within this definition.[2] Small states can be found in all corners of the world. However, most small states are found in the Pacific, the Caribbean, and the African Indian Ocean. Reflecting their global distribution, they are diverse in culture, geography, history, land area, levels of income, and economy. The majority of small states are island states. Some are isolated; others are landlocked or neighbours of much larger states. Many of them have been under colonial rule and have transplanted or mixed legal systems that reflect their colonial experience and heritage. A few are high-income countries; however, the majority are middle- or low-income countries. Some have fragile governance and are conflict-affected; others have lived under stable rule for centuries. Some small states are commodity exporters, while others have service and tourism-based economies.

Because of their size, small states face a set of common challenges including a vulnerability to external economic impacts such as changing trade regimes; many also have restricted ability to diversify their economic activity. They have generally limited public and private sector capacity. In particular, they face challenges in providing a complete legal and judicial infrastructure. They have an enhanced need for regional co-operation to combat any pressure international law and globalisation exert on them. Small island states are also particularly vulnerable to climate change.

---

[1]Operations Policy and Country Services, The World Bank (2016), p. ix. The series editors note that the definition is not uncontested. See Maass (2009), pp. 65–83.

[2]For the purpose of this series, the series editors include some territories within the definition of small states that are not classified as states as a matter of international law. These territories are geographically and culturally distinct entities that share the characteristics of small states, including the British Crown Dependencies of the Isle of Man, Jersey, and Guernsey and British Overseas Territories such as Gibraltar and the Pitcairn islands.

However, small states provide us with a unique opportunity to understand and gain insights not only into the experiences of larger states[3] but also, and more generally, concepts of governance, economics, cultural studies, sociology, and many other disciplines. They are often sites of social development and innovation since they are able to react more flexibly and more rapidly to challenges. They often have an influence in the world disproportionate to their size.

Despite the opportunities small states present in regard to the research and the study of pressing global problems, such as climate change, and also long-standing questions relating to ethics, legal pluralism, and colonialism, and international relations, scholarship, particularly legal scholarship, is relatively scarce. There are individual books on issues relating to small states,[4] and articles focusing on one or more small states can be found in general journals;[5] but small states research was in need of an interdisciplinary series devoted to showcasing and disseminating scholarship on small states. *The World of Small States* series, under the general editorship of Petra Butler and Caroline Morris, co-director of the Centre for Small States at Queen Mary University of London, is committed to publishing monographs and edited collections addressing small states issues in the areas of law, economics, politics, and international relations. We also welcome approaches from scholars in other disciplines.

The first volume of this series, *Small States in a Legal World*, is dedicated to some of the fundamental legal issues faced by small states. The volume begins with Geoffrey Palmer exploring the question whether a dystopian future for small island states and their unique culture can be avoided given the disproportional impact of climate change on them in 'Small Pacific Island States and the Catastrophe of Climate Change'.[6] The chapter explores that question by examining the likelihood of inundation from the sea and its consequences, by analysing whether the Paris Agreement will assist small island states and what the consequences of failure are. The chapter further discusses some of the human rights and the security issues that arise in regard to climate change in particular for small island states. The chapter is complemented by Alberto Costi and Nathan Jon Ross' chapter on 'The Ongoing Legal Status of Low-Lying States in the Climate-Changed Future'.[7] The authors discuss the consequences of climate change and the disappearance of small island states on the status of states under international law.

In Part II, *Small States: Challenges and Adventures in Law*, the reader can become acquainted with the diversity of issues that can be examined through the lens of the small state. In 'Competition Law and Policy in Small States', Lino

---

[3]Veenendaal and Corbett (2014), pp. 527–549.

[4]See, for example, Angelo (2014), Berry (2014), Briguglio (2014), Farran and Forsyth (2015), Corrin and Bamford (2015), and Thorhallsson (2000).

[5]It has to be noted some journals are, by virtue of their location, predominantly dealing with small states research and issues, such as the *Journal of South Pacific Law* or the *Caribbean Law Review*.

[6]Chapter 1, pp. 3–20.

[7]Chapter 6, pp. 101–138.

Briguglio of the Islands and Small States Institute shows, using Malta as a case study, that there are many factors associated with a small domestic market that have a bearing on competition law and policy.[8] Baldur Thorhallsson of the Centre for Small State Studies in 'Small States in the UNSC and the EU: Structural Weaknesses and Ability to Influence' investigates the methods and tools that small states can use to influence decision-making in the European Union and the United Nations Security Council.[9] 'The Impact of EU Law in Luxembourg: Does Size Matter?' by Michèle Finck[10] provides a case study for Thorhallsson's observations and an example, as Finck argues, of a state whose relationship with the EU can be viewed through the framework of size. Thorhallsson's observations are again tested in 'The Taxation of Small States and the Challenge of Commonality'[11] where Ann Mumford argues that by asserting a commonality of interest, smaller states may be able to perform beyond expectations in the international tax sphere and influence negotiations to the same extent as larger states. A common claim in studies of small states polities is that small size increases social cohesion and reduces the distance between citizens and their politicians. Therefore, small states should be model democracies. Derek O'Brien in his chapter 'Small States, Colonial Rule and Democracy' tests this perception by examining the state of democracy in the Caribbean and the reasons for it.[12] Tamasailau Suaalii-Sauni in 'Legal Pluralism and Politics in Samoa: The Faamatai, Monotaga and the Samoa Electoral Act 1963' examines the importance of being able to read cultural nuance in these socio-political reports and events and its relevance to understanding custom, the potential negative effects caused by the ambiguities created by the ad hoc blending of Samoa's fa'amatai (chiefly) and parliamentary democratic systems, and the lack of attention that theology has received in examinations of legal pluralism in the Pacific.[13]

The final part, *The Legal Profession in Small States: Education, Practice, and Regulation*, discusses aspects of the legal profession in the Pacific, Malta, Jersey, the Seychelles, and Cyprus. Trust in the legal system, in particular in its independent and ethical operation, is one of the cornerstones of a democratic state. In small societies, those issues are particularly pertinent. Since investment will more readily flow if a state has a robust legal profession and judiciary, the issues arising in regard to legal education and legal practice in small states are also one of economic development. The problems faced by the legal profession across the small states of the South Pacific and the various regulatory models that may be adopted to deal with ethical and other breaches of professional standards are examined by Nilesh Bilimoria in 'Choices for the South Pacific Region's Bar Associations and Law

---

[8]Chapter 2, pp. 23–34.

[9]Chapter 3, pp. 35–64.

[10]Chapter 4.

[11]Chapter 5.

[12]Chapter 7, pp. 139–163.

[13]Chapter 8, pp. 165–187.

Societies?'.[14] Nikitas Hatzimihail in 'On Law, Legal Elites and the Legal Profession in a (Biggish) Small State: Cyprus' explores the question of how size is impacting on the role—and functions—of law and lawyers in a small state that has a mixed legal system.[15] Seán Donlan, David Marrani, Mathilda Twomey, and David Zammit in 'Legal Education and the Profession in Three Mixed/Micro Jurisdictions: Malta, Jersey, and Seychelles' explore legal education and training and the legal profession in three mixed/micro jurisdictions: Malta, Jersey, and Seychelles.[16] The chapter considers how insiders in these jurisdictions look abroad to jurists and doctrine, judges and jurisprudence, and legislators and legislation, as well as foreign-trained practitioners, to orient their studies and practice. The effect of such external influences in small jurisdictions, the authors argue, is profound, especially in explicitly mixed traditions.

The general editors would like to thank all those involved in bringing the first volume of *The World of Small States* to fruition: the authors; the anonymous peer reviewers; Laura James and Niall Rand, LLB (Hons) graduates of Queen Mary University of London, who tirelessly and meticulously did a considerable part of the formatting, cite-checking, and additional research; and, finally, the team at Springer, especially Brigitte Reschke and Manuela Schwietzer, who responded enthusiastically to the series proposal and provided invaluable guidance along the way.

Wellington, New Zealand   Petra Butler
London, UK   Caroline Morris
November 2016

References

Angelo T (2014) Seychelles Digest. Law Publications, Seychelles

Berry D (2014) Caribbean integration law. Oxford University Press, Oxford

Briguglio L (ed) (2014) Building the resilience of small states: a revised framework. Commonwealth Secretariat, London

Corrin J, Bamford D (2015) Courts and civil procedure in the South Pacific. Intersentia, Cambridge

Farran S, Forsyth M (2015) Weaving Intellectual Property Policy in small island developing states. Intersentia, Cambridge

Maass M (2009) The elusive definition of the small state. Int Polit 46(1):65–83

Operations Policy and Country Services, The World Bank (2016) World Bank Group engagement with small states: taking stock. World Bank, Washington DC, p ix

Thorhallsson B (2000) The role of small states in the European Union. Ashgate, Aldershot

Veenendaal W, Corbett J (2014) Why small states offer important answers to large questions. Comp Polit Stud 48(4):527–549

---

[14]Chapter 11, pp. 245–264.

[15]Chapter 10, pp. 213–244.

[16]Chapter 9, pp. 191–212.

Contents

Part III The Legal Profession in Small States: Education, Practice and Regulation

About the Contributors

Nilesh Bilimoria is an assistant lecturer in the School of Law at the University of the South Pacific. He was formerly a senior legal officer with the Fiji Human Rights Commission and is currently a PhD candidate in the Faculty of Law at Queensland University of Technology.

Dr Lino Briguglio is professor of economics in the Faculty of Economics, Management and Accountancy at the University of Malta. He is the director of the Islands and Small States Institute and the Gozo Centre at the University of Malta. He has represented the Maltese government in many United Nations meetings, including the UN Global Conference on the Sustainable Development of Small Island States and the World Summit for Sustainable Development.

Alberto Costi is an associate professor in the Faculty of Law, Victoria University of Wellington, New Zealand. He is also the secretary-general of the International Law Association, New Zealand Branch, and vice president of the New Zealand Association for Comparative Law.

Dr Seán Donlan is associate professor and deputy head of school in the School of Law at the University of the South Pacific. He is a founding member of the micro jurisdictions research network *Nexus* and president emeritus of *Juris Diversitas*.

Dr Michèle Finck is a fellow in the Department of Law at the London School of Economics and a lecturer in EU law at Keble College, University of Oxford. She was formerly a visiting research scholar at New York University.

Dr Nikitas Hatzimihail is associate professor and vice-chair of the Department of Law at the University of Cyprus. He is also an affiliated senior researcher at the Université Libre de Bruxelles. He has served as advisor to the government of Greece on the Aegean and Islands Policy.

Dr David Marrani is the director of the Institute of Law in St Helier, Jersey, and a founding member of the micro jurisdictions research network *Nexus*.

Dr Ann Mumford is a reader in taxation law at the Dickson Poon School of Law at King's College London. She is a founding member of the *FemTax* network of international, feminist tax law professors.

Dr Derek O'Brien is a reader in law at the School of Law and a member of the Small Jurisdictions Service at Oxford Brookes University. He has served in the attorney general's chambers of the Cayman Islands and has advised the Foreign and Commonwealth Office and the Inter-American Development Bank on the provision of legal services in the British Overseas Territories and the independent countries of the Commonwealth Caribbean.

Sir Geoffrey Palmer QC is a distinguished fellow of the New Zealand Centre for Public Law and the Law Faculty at Victoria University of Wellington, New Zealand. He has served as attorney general, minister of justice, leader of the house, deputy prime minister, and prime minister of New Zealand. He was made a member of the Global 500 Roll of Honour by the United Nations Environment Programme. For eight years, he was New Zealand's commissioner to the International Whaling Commission.

Nathan Ross is a research fellow in international law and PhD candidate at Victoria University of Wellington, New Zealand, and a barrister and solicitor of the High Court of New Zealand.

Dr Tamasailau Suaalii-Sauni is an associate professor in the Department of Sociology at the University of Auckland, New Zealand. She regularly advises on Pasifika issues to numerous New Zealand bodies, including the Ministry of Health, the NZ Families Commission, and the Alcohol Advisory Council of NZ.

Dr Baldur Thorhallsson is professor and head of the Faculty of Political Science at the University of Iceland. He is also Jean Monnet chair in European Studies and Programme and research director at the Centre for Small States Studies at the University of Iceland.

Dr Mathilda Twomey is the chief justice of the Supreme Court of Seychelles. She is admitted to the Bar of England and Wales and practised as an attorney-at-law in Seychelles. She was a member of the Constitutional Commission of Seychelles which drafted the present Constitution of Seychelles in 1993.

Dr David Zammit is a senior lecturer and head of the Department of Civil Laws in the Faculty of Laws at the University of Malta. He is a fellow of the Royal Anthropological Institute and executive editor of the *Mediterranean Journal of Human Rights*.

Part I
2015 Keynote Lecture

Chapter 1
Small Pacific Island States and the Catastrophe of Climate Change

Geoffrey Palmer

1.1 Introduction

The international community decided at Paris in December 2015 to take some action to combat climate change. Given that the *United Nations Framework Convention on Climate Change* was agreed at the Earth Summit at Rio de Janeiro in 1992 and prior to Paris there were 20 Conferences of the Parties (COPs) to the Convention with little of substance achieved, one is entitled to ask whether the actions decided upon in 2015 are too little and too late. The twin pivots around which climate change policy revolves are mitigation and adaptation. To mitigate it is necessary to keep global warming by the end of this century to less than 1.5 °C and even then there will be adverse consequences, the increase in sea levels being particularly pertinent to small island states. If mitigation is not successful adaptation will have to do all the work. The consequences of climate change fall unevenly upon nations; some will fare better than others. Few will be worse affected than the small island states of the Pacific.

I will traverse the issues faced by small island states by exploring the following points:

- the vastness that is the Pacific
- the nations in the Pacific at greatest risk
- what the science says about the likelihood of inundation from the sea and its consequences
- what has the Paris agreement done to assist them?
- what are the consequences of failure?
- some discussion of human rights issues and the security issues.

G. Palmer (✉)
Faculty of Law, Victoria University of Wellington, Wellington, New Zealand
e-mail: geoffrey.palmer@vuw.ac.nz

© Springer International Publishing AG 2017
P. Butler, C. Morris (eds.), *Small States in a Legal World*,
The World of Small States 1, DOI 10.1007/978-3-319-39366-7_1

The issue with which I will leave you is this: can we avoid a dystopian future for Small Island States and their unique cultures?

1.2   The Pacific

When in Europe I am always struck by how small Europe is and how close the countries are to each other. It takes a New Zealander about 3 h to fly to eastern Australia (one of our closest neighbours). To Hawaii it is more than 9 h. To the United States mainland 12 h. And most places on the Pacific Rim take 10 h. The Pacific is one large ocean, covering a vast distance. So I will limit this inquiry to Oceania. By that expression I mean Australia, New Zealand, Papua New Guinea, Micronesia, Melanesia and Polynesia. Thirty million people live in Oceania. As you can see there are many islands. Oceania includes some of the smallest and most remote countries on the planet. They are not well known outside the region and not easily made the subject of global attention.

The Pacific Island Forum is the regional political association that covers many of these countries. Members include: Australia, Cook Island, Fiji, Kiribati, Marshall Islands, Federated States of Micronesia, Nauru, New Zealand, Niue, Palau, Papua New Guinea, Samoa, Tonga, Tuvalu, and Vanuatu. Associate Members include: New Caledonia, French Polynesia. Observers include: Tokelau, Wallis and Futuna, American Samoa, Timor Leste, Northern Mariana Islands.

1.3   Efforts by Vulnerable Nations

All nations are vulnerable to the threats from climate change but some are more vulnerable than others.

The Pacific Island Forum takes climate change seriously as an issue, more seriously than its most advanced members Australia and New Zealand have taken it so far. The most vulnerable countries have been critical of the stance taken by Australia and New Zealand on the issue, accusing them of not being real friends.[1] The Pacific Island Development Forum Summit headquartered in Fiji issued a very strong call for action on climate change a few days before the South Pacific Forum meeting.[2] The small island developing states know they are under the hammer despite having contributed very little to the cause of the climate change problem. In 2015 at Port Moresby Pacific Island Forum, leaders declared:[3]

---

[1]Australian Broadcasting Corporation (2015a, b).

[2]Pacific Islands Development Forum (2015a): This new grouping led by Fiji was formed because changing global and regional environment required new approaches to problem solving.

[3]Pacific Islands Forum Secretariat (2015b).

... that Pacific Island Countries and Territories are amongst the most vulnerable and least able to adapt and to respond; and the adverse consequences they face as a result of climate change, including the exacerbation of climate variability, sea level rise, ocean acidification, and more frequent and extreme weather events, are significantly disproportionate to negligible collective contribution to the global greenhouse gas emissions.

The Small Island Developing States (SIDS) of the world made concerted efforts to combine to bring maximum international pressure to bear long before the Paris meeting to ensure they were heard and their plight recognised. Their diplomatic efforts had success. In 2011 Palau's President urged the United Nations General Assembly to seek an advisory opinion from the International Court of Justice on whether states could be held liable for climate change under international customary law on transboundary harm.[4] The Alliance of Small Island States (AOSIS) had been active in the 20 previous COP meetings. The Maldives (in the Indian Ocean) was one of the founding members of this coalition of coastal and island nations highly vulnerable to climate change that was formed in 1990.

AOSIS states that it has a membership of 44 States and observers, drawn from all oceans and regions of the world: Africa, Caribbean, Indian Ocean, Mediterranean, Pacific and South China Sea. Thirty-nine are members of the United Nations, close to 28% of developing countries, and 20% of the UN's total membership. Together, SIDS communities constitute some 5% of the global population. Addressing the Security Council's Open debate on SIDS in 2015 the Secretary-General of the United Nations called for support of the nations in their actions to adapt to climate change.[5] The SIDS managed to secure much publicity in the run up to Paris and they certainly made a significant impact on the negotiations. AOSIS members at the Paris plenary started to sing Bob Marley's song 'Three Little Birds' repeating the refrain 'Every little thing gonna be all right' just before the final text was released; there was much cheering and applause.[6] Palau's ambassador to the EU and negotiator at Paris, Olai Uludong called the agreement 'remarkable'. The foreign minister of the Marshall Islands Tony de Brum spearheaded the High Ambition Coalition that secured support from 140 countries; 'it suddenly became the bus that everyone wanted to join' he said.[7] There were other expressions of relief from Pacific Island leaders at the result in Paris and the political achievement of these nations and the Paris agreement itself must be recognised and applauded after 20 failed attempts since 1992 at securing a meaningful international agreement. But just how meaningful Paris will be for these highly vulnerable nations must be analysed in a hard-headed way.

---

[4]Toribiong (2011).

[5]Ki-moon (2015).

[6]Little (2015).

[7]Pacific Islands Forum Secretariat (2015b).

1.4   What the Science Says

After years of prevarication and scepticism about the reality of climate change the science about it now seems to be widely accepted. In political terms acceptance that climate change is a reality that must be faced up to may be the greatest achievement of Paris. Yet the truth was clear enough from the time that the first report of the Intergovernmental Panel on Climate Change (IPCC) appeared in 1990.

As New Zealand's Minister for the Environment I made several speeches in the Pacific warning Pacific Island countries of the dangers of inundation from the sea due to rises in sea level attributed to climate change. I said at the University of Papua New Guinea in May 1989:[8]

> In our neighbourhood are many small nations, rich in history, culture and language. There are several nations in the Pacific region that are made up totally of atolls. The entire land base of these vital, unique and important countries may one day be physically destroyed.

The authoritative 2014 IPCC report summary for policymakers states the reality this way:[9]

> Anthropogenic greenhouse gas emissions have increased since the pre-industrial era, driven largely by economic and population growth, and are now higher than ever. This has led to atmospheric concentrations of carbon dioxide, methane and nitrous oxide that are unprecedented in at least the last 800,000 years. Their effects, together with those of other anthropogenic drivers, have been detected throughout the climate system and are extremely likely to have been the dominant cause of the observed warming since the mid-20th century.

Dr. James Hansen[10] from NASA and other leading scientists published a chapter in December 2013, the abstract of which says:

> Rapid emissions reduction is required to restore the Earth's energy balance and avoid ocean heat uptake that would practically guarantee irreversible effects. Continuation of high fossil fuel emissions, given current knowledge of the consequences, would be an act of extraordinarily witting intergenerational injustice. Responsible policymaking requires a rising price on carbon emissions that would preclude emissions from most remaining coals and unconventional fossil fuels and phase down emissions from conventional fossil fuels.

It is Hansen, with others, who has published the most recent research on sea level rise; in an article titled 'Ice melt, sea level rise and superstorms: evidence from paleoclimate data, climate modeling, and modern observations that 2 °C global warming is highly dangerous'.[11] In it, the authors explain how they found that temperatures were less than 1 °C warmer than today in prior interglacial periods (we are currently in an interglacial period, i.e. not an ice age) and that the sea level was 5–9 m higher than it is today. So, previous forecasts of sea level rise may be far

---

[8]Palmer (1990), p. 70.

[9]Intergovernmental Panel on Climate Change (2014a), p. 4.

[10]Hansen et al. (2013), p. 1.

[11]Hansen et al. (2016).

too conservative and the risks from global warming may be far greater than previously understood. Previous forecasts from the IPCC in 2013 predicted that sea levels will rise between half a metre and a metre by the end of this century. It seems clear from the most recent research, however, that these estimates will have to be revised upwards. The tipping point for icesheets in the Greenland ice sheet may arise at 1.6 °C above pre-industrial temperatures which is only 0.7 °C above today's temperatures.

I now wish to set out at length a draft of the problem definition on this subject by Nathan Jon Ross, a PhD student in law at the Victoria University of Wellington:

> The impacts from climate change on low-lying States are obviously acute. The common conception is "sinking islands". Due to thermal expansion of oceans and the melting of terrestrial ice and snow, sea level has risen and will continue to rise.[12] Indeed, the rate of sea level rise is increasing[13] and it is worse in the tropical Pacific, where many of the low lying States are situated, because the rate of increase in the region is up to four times higher than the global average.[14]

> However, the low-lying States are confronted by a much wider range of compounding climate change-related problems and it is worth pulling them together so readers have a more holistic understanding of their environmental situations.[15] Low-lying States are confronted by:

- Exacerbated weather extremes such as rainfall events and heat waves;[16]
- Flooding and inundation from sea-level rise and/or extreme weather events such as cyclones;[17]
- Marine water pollution and resulting salinisation of water supplies, agricultural lands and fresh water ecosystems;[18]
- Erosion of coastlines and coastal developments;[19]
- Bleaching and other damage to coral reefs, which compromises the ecosystem services they provide, for example, protecting island shores and providing habitat for marine species that are important to subsistence;[20] and

---

[12]Intergovernmental Panel on Climate Change (2013), p. 1139.

[13]Ibid; Hansen et al. (2015), p. 20059.

[14]Intergovernmental Panel on Climate Change (2014b), pp. 1619–1620; The variability in the extent of sea level rise is due to local geology and changes in ocean currents, see Fitzpatrick (2013).

[15]Intergovernmental Panel on Climate Change (2014b), pp. 1613–1642: Note that these problems are generalised here and that natural systems are complex and there are some naturally occurring processes that mitigate some of these events. However, the processes that mitigate these impacts are generally outstripped by the climate change effects that create these effects so that the overall effect is clearly negative.

[16]Intergovernmental Panel on Climate Change (2013); McLeman (2008), p. 11.

[17]Intergovernmental Panel on Climate Change (2014b), s 29.3.1.1.

[18]Barnett and Adger (2003).

[19]Intergovernmental Panel on Climate Change (2014b), s 29.3.1.1; McLeman (2008), p. 11.

[20]Intergovernmental Panel on Climate Change (2014b), s 29.3.1.2.

- Damaged mangroves and sea grasses caused by the greater depth of the sea, thereby compromising the ecosystem services they provide, such as providing foods, materials, and habitat for other species that are important for subsistence.[21]

From all of these environmental changes, there are human health impacts already observed, including:

- Direct mortality and injury from extreme weather events;[22]
- Increased incidences of diseases such as malaria and dengue fever;[23]
- Diseases from exposed landfill and burial sites following floods and inundation events;[24] and
- Compromised health from lack of access to freshwater and adequate nutrition.[25]

As developing countries, there are domestic issues that also contribute to challenges in these low-lying States, but the scale of these compounding environmental changes is clearly enormous and they simply do not have the capacity to protect themselves. Hence, the UN General Assembly has recognised that there may be instances where adaptation *in situ* is simply not feasible. The IPCC has noted that "it has been suggested that the very existence of some atoll nations is threatened by rising sea levels"[26] and that "land inundation due to sea-level rise poses risks to the territorial integrity of small-island states".[27]

There appears to be no definitive list of low-lying States that are the most at risk.[28] Although around 40 island nations face severe consequences,[29] the countries most often cited as being completely at risk are Tuvalu, Kiribati, the Marshall Islands and the Maldives.

Looking at the situation of these countries more closely I suggest we need to ask whether they can realistically be saved from the effects of climate change by any combination of mitigation and adaptation.

The highest points on these island nations are:

- *Kiribati*: 81 m.
  Kiribati comprises 1 island (Banaba) and 32 atolls. The highest point is on the island of Banaba, which is just 6 km long, and has been destroyed by phosphate mining. It supports a population of just 300. The original Banaban people now live on the territory of Fiji. In the atolls, where the remaining 112,000 people live, the highest point is just 2 m.
- *Tuvalu*: 4.6 m.

---

[21]Ibid.

[22]Intergovernmental Panel on Climate Change (2014b), s 29.3.3.2.

[23]Ibid.

[24]Foster (2014).

[25]Intergovernmental Panel on Climate Change (2014b), s 29.3.3.2.

[26]Intergovernmental Panel on Climate Change (2014b), Chap. 29.

[27]Intergovernmental Panel on Climate Change (2014b), Chap. 21.

[28]Park (2011), p. 2.

[29]Warner et al. (2009), s 3.7.

There are three reef islands and six atolls. The population is 10,837.
- *Marshall Islands*: 10 m.
  There are five islands and 29 atolls. The population is 56,719.
- *The Maldives*: 2.4 m, the lowest 'highest point' in the world.
  There are 26 atolls that make up the Maldives. The population is 402,071. The Maldives are situated in the Indian Ocean.

I should note here that Tokelau comprises three atolls north of Samoa just 3–5 m above sea level, with a population of about 1400. About 9000 Tokelauans live in New Zealand now. Tokelau is part of the territory of New Zealand. So the people are New Zealand citizens and should be able to relocate. New Zealand provides assistance to Tokelau. But climate change challenges the future of Tokelau. There has been little discussion in New Zealand about this that I have been able to discern.

Although the height above sea level is important context, there is more to the story than just a consideration of this height above sea level, as indicated by the example of 81 m high Banaba Island being virtually uninhabitable. The wide range of environmental and social consequences of climate change, the idea of 'sinking states' is just one of the issues.[30] The key point is this: even if some *terra firma* remains with its head above water, it will not be in any condition to support a thriving human community. Ex situ adaptation is almost certainly inevitable.

Can these island nations be saved via a combination of mitigation and adaptation? According to the research by Hansen et al., it seems that sea-level rise—without any of the other adverse effects—will be enough to cause these countries to become completely uninhabitable. Kiribati President Anote Tong has already conceded that some of the atolls are already destined to be lost and hence runs a government policy called 'migration with dignity', which seeks to relocate people in a way that is conducive to community and positive contribution to the host country.[31]

The costs of in situ adaptation seem to be completely prohibitive. My colleague at Victoria University of Wellington, Alberto Costi, has published an article on small island states and statehood in which he concluded that:[32]

> While there are human-made options to prevent the effects of climate change on low lying atoll states, these are unrealistic and unaffordable. The Maldives has looked into island protection, but costs of US$6 billion for coastal protection, or US$500–1,000 million to elevate islands by one meter, were deemed too expensive. Atoll states in the Pacific, relying on an annual gross domestic product (GDP) ranging from US$27 million in Tuvalu to US$644 million in Vanuatu, would be similarly unable to afford such measures. In the current state of play, the disappearance of low lying atoll states is unlikely to be prevented through human-made techniques.

A more extreme solution proposed by some is to construct artificial islands. There are many engineering, environmental, social and cultural reasons why this

---

[30]Gerrard and Wannier (2015).

[31]Office of the President of Kiribati (2016).

[32]Costi (2014), p. 145.

option is not practical, but ultimately the astronomical cost will be a barrier. The idea that developing countries (even accounting for international aid) could construct artificial islands that are satisfactory for homes, economies and subsistence for hundreds of thousands of islanders appears to be in the realm of science fiction, particularly when contrasted with the relative ease of moving to other territory.

It is important to note that a great many countries will have serious problems of adaptation to face with inundations from the sea, it is widely recognised by scientists in New Zealand that this will be the most serious of all climate change issues for New Zealand.[33]

1.5 What Has the Paris Agreement Done for SIDS?

The political achievement at Paris was substantial, the legal achievement much less.

The Paris Agreement is long on aspiration and short on obligation. The negotiating strategy devised for Paris called for nations to make Intended Nationally Determined Contributions (INDCs).These were not intended to be and are not legally binding. As expected, the cumulative offers received at Paris fell well short of what will be required to keep the temperature below 2 °C by the end of the century, let alone 1.5 °C.

Since the objective of the 1992 Framework Convention on Climate Change is stabilisation of greenhouse gas concentrations in the atmosphere at a level that would prevent dangerous anthropogenic interference with the climate system, we obviously have a long way to go. We are not there yet or anywhere near. So in those terms Paris is not a success. The absence of binding targets on nations means there can be no effective enforceability of the INDC commitments, inadequate though they are.

This approach was deliberate on the part of negotiators, trying to avoid some of the traps that the Kyoto Protocol fell into. They tried to keep the developing countries in the tent and to accommodate the United States, where the prospect of securing Senate consent for binding targets looked hopeless, largely due to Republican party attitudes concerning climate change denial.

Thus, the issue is whether the Paris Agreement will after further iterations ripen into a success and achieve mitigation to the level required within the time available. That in turn will depend upon how far the political momentum generated at Paris will continue in order to ultimately produce sufficient binding obligations. The calculation was that an agreement with everyone on board was better than one where they were not, even if the price paid was to lower the level of ambition. The Paris Agreement can best be understood as a global commitment to a future continuing process to address climate change issues.

---

[33]Meduna (2015), pp. 34–35.

Lawyers deal in binding obligations. In order to find these, they look closely at the text. The text arising from Paris speaks with several voices. Analysing the binding obligations flowing from the negotiations and those which are aspirational may help in assessing the achievement. The Paris Agreement has some binding elements. And there are some binding elements of the Framework Convention and later instruments but many of these are not directly relevant to mitigation. And at this point mitigation is the critical issue.

Given the difficulties facing the negotiators and the failures of the past, the legal architecture of the Paris Agreement has impressive and interesting elements. Clearly, the strategic aim was to pull all nations into the Agreement by being very inclusive and avoiding division and confrontation. The strategy reminds me of the old nursery rhyme, 'Come into my parlour said the spider to the fly'. Once caught in the web the tentacles of the agreement will tighten later and it may be very difficult for nations to remove themselves because they would be likely to lose a lot of face. Shaming in its various forms remains one of the most potent of international sanctions.

Let me begin my analysis by starting at the end. Article 27 of the Agreement permits no reservations to it to be made. Nations can withdraw after 3 years from the date the Agreement entered into force. And the Agreement enters into force on the thirtieth day after the date 'on which at least 55 parties to the Convention accounting in total for at least an estimated 55 percent of the total global greenhouse gas emissions having deposited their instruments of ratification, acceptance, approval or accession'.[34]

The agreement is open for signature from 22 April 2016 to 21 April 2017. It is open for accession from the day following the date upon which it is closed for signature. So it will be a long time before we will know who has signed up and who has ratified. And ratification is the act 'whereby a State establishes on the international plane its consent to be bound by a treaty'.[35] I judge it will be 2018 at the earliest before we receive the necessary ratification answers and can therefore analyse what the precise legal effect of the Agreement is. It does not seem in this case that signature alone will be sufficient for a State to be legally bound.

A cunning feature of the Agreement is that a great deal of activity will take place within the councils of the Convention system before ratification occurs. Much detailed and specialised machinery was set running in Paris and while much of this does not involve legal obligations imposed on States it does mean that a great deal of work will rapidly occur that is likely to make the nature of future decisions clearer and possibly easier for States to swallow.

The forward momentum is achieved by the bifurcated nature of the agreement. It comes in two parts. In a total package of 31 pages of text only 11 pages constitute the binding Paris Agreement. It is preceded by 19 pages of 'decisions' made by the COP. It was decided to adopt the Paris Agreement under the UNFCC although the

---

[34]United Nations Framework Convention on Climate Change (2015), art 21.

[35]United Nations Treaty Series (1969), art 2.

relationship between the Paris Agreement and the Convention is not on all fours. Some nations may adhere to one and not the other.

The dispute settlement mechanism for the Paris Agreement is the same as for the Convention itself. Disputes are likely to occur between nations and the effectiveness of the Agreement may depend on how efficient the method is in settling disputes. Over time the emphasis is likely to move to enforceability issues. Nations are enjoined to seek settlement of a dispute through negotiation or any other peaceful means of their own choice.[36] That mechanism stipulates that nations when ratifying the Agreement may state 'in respect of any dispute concerning the interpretation for application . . . it recognizes as compulsory ipso facto and without special agreement, in relation to any party accepting the same obligation . . . submission to the International Court of Justice and/or Arbitration'. Such declarations are not mandatory. Where the process does not produce a resolution after 12 months, the dispute is submitted to a conciliation commission, created upon the request of one of the parties. While this dispute settlement mechanism is not as strong as domestic law enforcement through municipal Courts, it is stronger than many international environmental treaties.

The first part of the document recording decisions of the COP also records the decision to establish an Ad Hoc Working Group on the Paris Agreement. That group will prepare for entry into force of the Agreement and oversee the implementation of a work programme. This will start in 2016.

There is a great deal in the non-binding text about INDCs and it notes that the 'much greater emission reduction efforts will be required in order to hold the increase in global average temperature to below 2 °C above pre-industrial levels by reducing emissions to 40 gigatonnes or to 1.5 °C above pre-industrial levels'. The COP also decided to invite the IPCC to provide a special report in 2018 on the impacts of global warming of 1.5 °C above pre-industrial levels and related global gas emissions pathways. The language in this part of the text evinces an intention to ratchet up the INDCs. The language requires parties to submit their INDC at least 9–12 months before the relevant COP. There is emphasis on both clarity and transparency. And there will be guidance for accounting of INDCs. Work in abundance is also ordered up from the Subsidiary Body for Scientific and Technological Advice. Other Committees of Expert groups are loaded with work. The Adaptation Committee also receives instructions.

Finance is the subject of heavy attention and more work. It is the same for technology development and transfer. Capacity building similarly for developing countries also receives a big work plan. There is also a capacity building initiative for transparency to build institutional and technical capacity to advance Article 13 of the Agreement. There is also activity around facilitating implementation and compliance. Resolution was also made to enhance the provision of urgent and adequate finance with a roadmap to be produced to secure USD $100 billion annually by 2020 for mitigation and adaptation.

---

[36]United Nations Framework Convention on Climate Change (2015), art 14(2).

Much of the language in the 19 pages of COP decisions talks of 'requesting', 'encouraging', 'striving' and similar hortatory language that speaks not the language of State obligation. But there emanates from the document a sense of urgency and vigorous activity on a wide range of fronts. The work seems designed to advance the Agreement itself in quite a rapid way. Plenty will happen quickly, as it needs to do.

All of this reflects the political break-through that Paris achieved. And much of the work is explicitly aimed at providing help to developing countries of a practical and useful sort. But activity, however well directed and however productive, is not in itself a substitute for binding legal obligations. Nevertheless, the flavour of Paris is to work towards binding legal obligations that will limit emissions. The first 19 pages of text were important in the sense they engender a feeling of activity. The bare features of the legal Agreement itself would have looked very thin without the COP decisions. The very lengthy 'Decisions to give effect to the Agreement' gives the impression of substantial even frenetic activity. Halting climate change could be the outcome in the future.

As for the Agreement itself, Article 2 provides there is a commitment to 'holding the increase in global average temperature to well below 2 °C above pre-industrial levels and to pursue efforts to limit the temperature increase to 1.5 °C above pre-industrial levels, recognizing that this would significantly reduce the risks and impacts of climate change'. There are similar commitments to increasing the ability to adapt to climate change and foster climate resilience and making finance flow 'consistent with a pathway towards low greenhouse gas emissions and climate-resilient development'. This general type of commitment can hardly be said to create any binding obligations on the States themselves. Article 2 is not a hard commitment to hold the increase to 1.5 °C and we do not know what 'well below 2 °C' may mean. There is a commitment to pursue efforts to hold the temperature increase to 1.5 degrees but that is not an obligation to achieve it. But all States are require to 'undertake and communicate ambitious efforts' as defined in the Agreement 'with a view to achieving the purpose of the agreement'. And this process will be a progression over time.

The agreement itself comes closer to imposing specific obligations on States than the decisions of the COP, but many of the Articles contain principles rather than specific obligations. Nevertheless, there are hard law obligations to report and communicate about various matters such as each party being required to account 'for their nationally determined contributions', as required by Article 4, in quite defined ways. But in many of these Articles there remain large elements of discretion left to States, and gaps. Language such as 'should', 'flexibility', 'strive' and 'aim' and all the familiar weasel words of international agreements are employed.

Article 4 states '... parties aim to reach global peaking of greenhouse gas emissions as soon as possible ... and to undertake rapid reductions thereafter ... so as to achieve a balance between anthropogenic emissions by sources and removals by sinks of greenhouse gases in the second half of this century ...'. This language contains an important commitment, the precise nature of which is susceptible to a number of interpretations. What does 'balance' mean? It does not say 'net zero emissions' but does it mean that? And when will 'peaking' occur?

There is imposed a legal duty on parties to account for their emissions and to prepare every 5 years and communicate successive NDICs to reflect its highest ambition and each successive one 'will represent a progression' beyond the party's then INDC. (How binding this requirement will be is not easy to judge. What happens if nations fail to comply?) The special circumstances of developing countries and small island developing States is recognised in this Article. Accounting for INDCs is mandatory and parties 'shall promote environmental integrity, transparency, accuracy, completeness comparability and consistency, and ensure the avoidance of double counting'.[37] Each party to the Agreement is responsible for its emissions levels. Countries are obliged to pursue policies with the aim of achieving their pledges. As far as I can see there is no legal obligation on nations, although the Decisions text invites them to do so. The Agreement says they are to 'strive' to write a low emissions strategy by 2020.

Parties "should take action to conserve and enhance carbon sinks and reservoirs of GHGs, with an emphasis on forests. Parties 'are encouraged' to reduce emissions from deforestation.[38]

Parties are free to choose voluntary cooperation in implementation of INDCs with cooperative approaches that involve the use of internationally transferred mitigation outcomes. A mechanism is established by the Agreement to facilitate the market. It will be supervised by a body designated by the COP. Rules and procedures will have to be adopted. At the same time integrated, holistic and balanced non-market approaches and a framework to promote these are established by the Agreement, but without any detail. No mechanism has been set up to set an international carbon price, but it is possible a club approach by big emitters agreeing amongst themselves could cause such a price to emerge.

Adaptation is advanced by agreement on the 'global goal on adaptation of enhancing adaptive capacity strengthening resilience and reducing vulnerability to climate change'. Each party shall 'as appropriate engage in adaptation planning processes and the implementation of actions, including plans and policies'. There is encouragement to strengthen cooperation. Parties 'should' submit and update periodically an adaptation communication setting out its plan, actions and priorities. It is to be recorded in a public registry. There is nothing much here in the nature of hard law obligations.

Article 8 on loss and damage revolves around the Warsaw International Mechanism for Loss and Damage associated with climate change and this may be enhanced and strengthened. It is far from clear what will come out of the loss and damage work. But it may be a positive move that it has been separated from measures for adaptation to climate change. But it is made clear in the Decisions

---

[37] Ibid, art 4(13).

[38] Ibid, art 5.

of the COP that the Agreement itself does not involve or provide a basis for any liability or compensation for loss and damage.

Article 9 deals with financial resources to assist developing countries; 'developed country parties shall provide financial resources to assist developing country parties with respect to both mitigation and adaptation in continuation of their existing obligations under the Convention'. That is a legal duty but it is very generalised and lacks the specificity required for enforcement.

Article 10 is concerned with technology development and transfer; 'accelerating, encouraging and enabling innovation is critical for an effective, long-term global response to climate change and promoting economic growth and sustainable development'.

On capacity building Article 11 of the Agreement says capacity building 'should enhance the capacity and ability of developing countries ... and those that are particularly vulnerable to the adverse effects of climate change, such as small island developing states to take effective climate change action ...'. No enforceable legal obligations arise here.

Article 12 is succinct. It erects a legal duty of cooperation 'to enhance climate change education, training, public awareness, public participation and public access to information, recognizing the importance of these steps with respect to enhancing actions under this Agreement'.

If this Agreement solves the problem of climate change in the future it will be, in my opinion, because of the provisions in Articles 13 and 14 operating in conjunction with Article 4. Enhanced transparency is the goal of Article 13, which provides that 'an enhanced transparency framework for action and support, with built-in flexibility which takes into account parties' different capacities and builds upon collective experiences is hereby established'. Its purpose is 'to provide a clear understanding of climate change action in light of the objective of the Convention as set out in its Article 2, including clarity and tracking of progress towards achieving parties' individual nationally determined contributions under Article 4, and parties' adaptation actions under Article 7 including good practices, priorities, needs, and gaps to inform the global stocktake under Article 14'.

There is a transparency of action and a transparency of support—both are established. The reporting requirements established for both of these is specific and is the subject of technical expert review. Provision of the required information and reporting is mandatory. This appears to be one of the most effective provisions in the Agreement and one likely to make a difference. A periodic global stocktake of the implementation of the Agreement is required every 5 years, the first being in 2023. This should be of material assistance in reaching the goal of the Agreement.

Article 15 provides for a mechanism to facilitate implementation and to promote compliance with the Agreement. An expert-based committee is established for this purpose. It must function 'in a manner that is transparent, non-adversarial and non-punitive'. Its procedures remain to be settled.

It is difficult to assess how all this will work or whether it will produce the desired outcome in time. It all depends upon continuing political will. And there are many geo-political problems that could knock the process all off-course. But it has to be said it is a positive start, if a long delayed one. The hallmark of all the previous 20 COPs has been procrastination, putting off hard decisions until later. That has occurred again here and time is rapidly running out. An earlier start would have made the problems of adjustment much easier. But it is a start. The mechanism for ratcheting up the INDCs over time may prove to be a successful strategy. But it must be understood the rate of decarbonising the economies of the world must be very rapid and there is nothing in the Agreement that directly addresses fossil fuels and their use, only the emissions that result from that use.

The Agreement has potential. Whether the potential will be realised is a question of speculation. The Agreement does not assure us that temperatures will not rise by more than 2 °C, let alone be restricted to 1.5 °C. It is more a political agreement to keep trying than a legal set of binding obligations that will produce the necessary result. But the Agreement does contain sufficient binding obligations that could in time make a difference. When the detail has been developed we will know more.

I am not prepared to judge at this juncture whether the approach taken in Paris will succeed. I hope it does. But I seriously doubt that it will succeed in holding the warming to 1.5 °C by 2050. The next test will be the Climate Summit in Morocco in November 2016. Then in 2020 the emission cutting plans of nations must be submitted. Paris could succeed or it could fail as Kyoto did. For me it is a case of one hand clapping. Whether the Paris Agreement will succeed in combatting climate change is a question with no answer at this juncture.

The Paris Agreement does offer some solace to SIDS but it appears to be too little too late. The reference to 1.5 °C is directed at their interests. They succeeded in having that reference included in the Agreement. They will secure funding from the Green Climate Fund to assist them in adaptation. There will be more funds made available and they will be better directed than in the past. There will be more research into and enhanced transfer of technologies that may assist them. More technical support is likely to be made available. They will secure help to develop their capacity to cope with the problems as explicitly provided for in Article 11. They will be cut some slack in complying with aspects of the Agreement. They will benefit from the sense of urgency that infuses the whole international effort. However, in terms of the four nations whose plight was analysed above it seems difficult to conclude that they will be saved from a level of sea level rise that will endanger their existence and their future as nations.

1.6   Human Rights, Security and People Movement

In 2009 the United Nations Commissioner for Human Rights produced a report upon the consequences of projected climate change upon the enjoyment of human rights. Its conclusions make sobering reading:[39]

---

[39]United Nations Human Rights Council (2009).

Climate change-related impacts, as set out in the assessment reports of the Intergovern-mental Panel on Climate Change, have a range of implications for the effective enjoyment of human rights. The effects on human rights can be of a direct nature, such as the threat extreme weather events may pose to the right to life, but will often have an indirect and gradual effect on human rights, such as increasing stress on health systems and vulnera-bilities related to climate change-induced migration.

The effects of climate change are already being felt by individuals and communities around the world. Particularly vulnerable are those living on the "front line" of climate change, in places where even small climatic changes can have catastrophic consequences for lives and livelihoods. Vulnerability due to geography is often compounded by a low capacity to adapt, rendering many of the poorest countries and communities particularly vulnerable to the effects of climate change.

Within countries, existing vulnerabilities are exacerbated by the effects of climate change. Groups such as children, women, the elderly and persons with disabilities are often particularly vulnerable to the adverse effects of climate change on the enjoyment of their human rights. The application of a human rights approach in preventing and responding to the effects of climate change serves to empower individuals and groups, who should be perceived as active agents of change and not as passive victims.

Often the effects of climate change on human rights are determined by non-climatic factors, including discrimination and unequal power relationships. This underlines the importance of addressing human rights threats posed by climate change through adequate policies and measures which are coherent with overall human rights objectives. Human rights standards and principles should inform and strengthen policy measures in the area of climate change.

The physical impacts of global warming cannot easily be classified as human rights violations, not least because climate change-related harm often cannot clearly be attributed to acts or omissions of specific States. Yet, addressing that harm remains a critical human rights concern and obligation under international law. Hence, legal protection remains relevant as a safeguard against climate change-related risks and infringements of human rights resulting from policies and measures taken at the national level to address climate change.

There is a need for more detailed studies and data collection at country level in order to assess the human rights impact of climate change-related phenomena and of policies and measures adopted to address climate change. In this regard, States could usefully provide information on measures to assess and address vulnerabilities and impacts related to climate change as they affect individuals and groups, in reporting to the United Nations human rights treaty monitoring bodies and the United Nations Framework Convention on Climate Change.

Further study is also needed of protection mechanisms for persons who may be considered to have been displaced within or across national borders due to climate change-related events and for those populations which may be permanently displaced as a consequence of inundation of low-lying areas and island States.

Global warming can only be dealt with through cooperation by all members of the international community. Equally, international assistance is required to ensure sustainable development pathways in developing countries and enable them to adapt to now unavoid-able climate change. International human rights law complements the United Nations Framework Convention on Climate Change by underlining that international cooperation is not only expedient but also a human rights obligation and that its central objective is the realization of human rights.

The analysis contained in the report all seems obvious enough and has been reinforced by the later findings of the Intergovernmental Panel on Climate Change. The trouble with all the analyses and reports that have been produced over the years on the subject of climate change is that political leaders either do not read them or do not heed them. The immediacy of the 'politics of now' leads them to be reckless about the future and to put decisions off for the next generation of leaders. We are in 2016 dangerously close to the point where time is running out and the longer we leave it the harder the adjustments will be.

Over time climate change is likely to have serious implications for international peace and security. The movements of large numbers of people from low lying areas will produce unrest unless it is carried out in an orderly and planned manner. The issue has been raised in the Security Council on four occasions, most recently in June of 2015.[40] But the Security Council has shied away from the climate change issue, preferring to leave it to the UN Framework Convention process. One would have thought the threats to the integrity of low lying island states should be sufficient to engage the interest of the Security Council. China, Russia and India and many developing countries have been reluctant to engage the Security Council apparently because the Council does not operate under the principles of common but differentiated responsibility which underpins important aspects of the United Nations climate change negotiations.

The consequences of climate change for the SIDS will be serious post-Paris. While they may not be totally underwater their ability to support their populations will be seriously impaired, there will be massive disruption and inconvenience and great pressure on governments to provide homes elsewhere for their people. It is quite likely that the future of these nations is jeopardised and they may cease to exist Their cultures, their identity and their history may be brought to an end. What a tragedy that would be.

Post-Paris the international community needs to come up with carefully researched but bold policies concerning how to handle these foreseeable problems that will not be prevented by the Paris Agreement. A dystopian future for these communities needs to be avoided. It can be avoided by determined international actions.

References

Australian Broadcasting Corporation (2015a) Marshall Islands foreign minister Tony de Brum slams Australia's 2030 carbon emissions targets. Available at http://www.abc.net.au/news/ 2015-08-11/marshall-islands-slams-australia's-carbon-emissions-targets/6688974. Accessed 28 June 2016

Australian Broadcasting Corporation (2015b) Pacific Islands forum: Kiribati urges Australia, NZ to be 'real friends' on climate change. Available at http://www.abc.net.au/news/2015-09-07/ Kiribabi-urges-australia-nz-to-be-real-friends-on-climate-change/6755794. Accessed 28 June 2016

Barnett J, Adger WN (2003) Climate dangers and atoll countries. Climate Change 61(3):321–337

---

[40]United Nations (2015).

Costi A (2014) Climate change and the legal status of a disappearing state in international law. Int Law Read 12:140–177

Fitzpatrick M (2013) What accounts for the varying rates of sea level rise in different locations? Union of Concerned Scientists. Available via http://www.ucsusa.org/publications/ask/2013/sea-level-rise.html. Accessed 28 June 2016

Foster J (2014) Epic king tides offer glimpse of climate change in Marshall Islands. Available at http://thinkprogress.org/climate/2014/03/06/3372301/marshall-islands-flood-king-tide. Accessed 28 June 2016

Gerrard MB, Wannier GE (eds) (2015) Threatened island nations: legal implications of rising seas and a changing climate. Cambridge University Press, New York

Hansen J, Kharecha P et al (2013) Assessing "dangerous climate change": required reduction of carbon emissions to protect young people, future generations and nature. PLoS One 8(12):1–26 Available at http://dx.doi.org/10.1371/journal.pone.0081648. Accessed 28 June 2016

Hansen J et al (2015) Ice melt, sea level rise and superstorms. Atmos Chem Phys Discuss 15:20059–20179

Hansen J, Sato M, et al (2016) Ice melt, sea level rise and superstorms: evidence from paleoclimate data, climate modeling, and modern observations that 2°C global warming is highly dangerous. Available at http://www.atmos-chem-phys.net/16/3761/2016/acp-16-3761-2016-discussion.html. Accessed 28 June 2016

Intergovernmental Panel on Climate Change (2013) Climate Change 2013: The physical science basis. In: Stocker TF, Qin D, Plattner G-K, Tignor M, Allen SK, Boschung J, Nauels A, Xia Y, Bex V, Midgley PM (eds) Contribution of Working Group I to the Fifth assessment report of the intergovernmental panel on climate change. Cambridge University Press, Cambridge, p 1535

Intergovernmental Panel on Climate Change (2014a) Climate Change 2014: Synthesis Report, Summary for Polcymakers, Contribution of Working Groups I, II and III to the Fifth Assessment Report of the Intergovernmental Panel on Climate Change. In: Core Writing Team, Pachauri RK, Meyer LA (eds) IPCC, Geneva, Switzerland

Intergovernmental Panel on Climate Change (2014b) Climate Change 2014: Impacts, Adaptation, and Vulnerability. Part B: Regional Aspects. Contribution of Working Group II to the Fifth Assessment Report of the Intergovernmental Panel on Climate Change. In: Barros VR, Field CB, Dokken DJ, Mastrandrea MD, Mach KJ, Bilir TE, Chatterjee M, Ebi KL, Estrada YO, Genova RC, Girma B, Kissel ES, Levy AN, MacCracken S, Mastrandrea PR, White LL (eds) Cambridge University Press, Cambridge, p 688

Ki-moon B (2015) Addressing the Security Council's open debate on small island developing states, Secretary General urges global push to combat crime, climate change threats. Available at http://www.un.org/press/en/2015/sgsm16981.doc.htm. Accessed 27 June 2016

Little A (2015) What the Paris Climate Change Agreement means for vulnerable nations. The New Yorker. Available at http://www.newyorker.com/news/news-desk/what-the-paris-climate-agreement-means-for-vulnerable-nations. Accessed 28 June 2016

McLeman R (2008) Climate change migration, refugee protection, and adaptive capacity building. McGill Int J Sustain Dev Law Policy 4(1):1–18

Meduna V (2015) Towards a warmer world: what climate change will mean for New Zealand's future. Bridget Williams Books, New Zealand

Office of the President of Kiribati (2016) Relocation. Available at http://www.climate.gov.ki/category/action/relocation. Accessed 28 June 2016

Pacific Islands Development Forum Secretariat (2015a) Suva Declaration on Climate Change from the Third Annual Summit held in Suva, Fiji 2–5 September 2015

Pacific Islands Forum Secretariat (2015b) Pacific Island Forum Leaders Declaration on Climate Change Action from the 46th Pacific Island Forum Leaders Summit held in Port Moresby, Papua New Guinea 8–10 September 2015

Palmer G (1990) Environmental politics. Dunedin, New Zealand

Park S (2011) Climate change and the risk of statelessness: the situation of low-lying island states. The United Nations High Commissioner for refugees, legal and protection policy research series

Toribiong J (2011) Statement by the Honourable Johnson Toribiong, President of the Republic of Palau to the 66th Regular Session of the United Nations General Assembly on 22 September 2011. Available at www.gadebate.un.org. Accessed 28 June 2016

United Nations (2015) Meeting coverage: issues facing small island developing states 'global challenges' demanding collective responsibility, Secretary-General tells Security Council. Available at http://www.un.org/press/en/2015/sc11991.doc.htm. Accessed 28 June 2016

United Nations Framework Convention on Climate Change (2015) Paris Agreement

United Nations Human Rights Council (2009) Report of the Office of the United Nations High Commissioner for Human Rights on the relationship between climate change and human rights. UN Doc A/HRC/10/61

United Nations Treaty Series (1969) Vienna Convention on the Law of Treaties (with annex). 1155 UNTS 332

Warner K, Erhart C et al (2009) In search of shelter: mapping the effects of climate change on human migration and displacement. CARE International

Part II
Small States: Challenges and Adventures in Law

Chapter 2
Competition Law and Policy in Small States

Lino Briguglio

2.1 Introduction

This chapter will attempt to show that there are many factors associated with a small domestic market that have a bearing on competition law and policy, and therefore, the competition regimes of a small state should take these factors into account. Special reference will be made to Malta, where competition legislation is modelled on EU competition law.

The chapter is organised as follows: Sect. 2.2, which follows this introduction, lists the characteristics of small states which may have a bearing on competition law and policy; Sect. 2.3 the implications for the implementation of competition law, associated with a small domestic market; Sect. 2.4 concludes the study by summarising the main thrust of the arguments put forward in this chapter.

2.2 Characteristics of Small States

2.2.1 What Is a 'Small State'?

The size of a country can be measured in terms of its population, its land area or its gross domestic product.[1] Some studies prefer to use population size as an index of country size, while others take a composite index of the three variables. There is no

---

This chapter is an updated and revised version of the study titled 'Competition constraints in small jurisdictions' published in Bank of Valletta Review number 30 (2004). The author would like to thank Dr Sylvann Aquilina Zahra for her comments on an earlier draft of this chapter.

[1]On this question see Downes (1988); Jalan (1982); Briguglio (1995).

L. Briguglio (✉)
Faculty of Economics, Management and Accountancy, University of Malta, Msida, Malta
e-mail: lino.briguglio@um.edu.mt

© Springer International Publishing AG 2017
P. Butler, C. Morris (eds.), *Small States in a Legal World*,
The World of Small States 1, DOI 10.1007/978-3-319-39366-7_2

general acceptance as to what constitutes a small state, although a state with a population of around 1.5 million or less would generally be considered as a small one.[2]

So far there has not been any attempt to classify countries according to the size of their domestic market, although the issue has been discussed in a few studies.[3] One possible indicator could be a composite index consisting of population multiplied by real consumption expenditure, suitably standardised for international comparisons. Such an index would take account of the number of actors and the value of transactions within a given market. A cut off point would also be needed to establish whether a domestic market, in a given state, is to be considered as a small one.

2.2.2   Small Domestic Market

Small states are likely to have a small domestic market, which in turn limits competition possibilities due to the ease of market dominance by firms.

In addition, a small domestic market tends to be characterised by natural monopolies in utilities, such as electricity, fixed line telephony, gas, and water, where the relatively large overhead costs, with the existing technology, do not permit more than one entity to supply the service viably.

Another characteristic of small markets relates to barriers to entry. There are natural barriers, due to the poor chances of success of getting new business in goods and services already supplied by existing firms. In addition, in a small market, bulk buying is often required to avoid excessive fragmentation of cargoes, especially in the case of raw materials, and this limits the number of players in that market. There may also be artificial barriers to entry, often imposed by governments, to make it viable for a business to invest in certain types of production of goods and services, where overhead costs are large, and hefty capital outlays are required. In many cases, entry is also limited in the provision of services where competition could be possible, but the nature of the service requires licensing.

Still another characteristic of small states is parallel behaviour between firms, which tends to be easier to conduct where family ties predominate in business. In such circumstances, the competition authorities may find it difficult to distinguish between concerted practices and independent action.[4] In addition, arrangements between importers and distributors involving restrictions with the aim of minimising intra-brand competition may be easier to put in place in small states. Although this is likely to stem from self-interest, it may have beneficial impacts on

---

[2]Commonwealth Secretariat (2016): This is the population threshold adopted by the Commonwealth Secretariat.

[3]See for example Armstong and Read (1998); Murphy and Smith (1999); Gal (2002); International Competition Network (ICN) (2009); McKoy (2009).

[4]Muscat (1998).

the consumer since uncontrolled competition may usher in excessive fragmentation and instability. This issue will be discussed further below.

The size of the population has a direct bearing on the running and enforcement of competition law and policy. In small states, the chances of finding the necessary expertise to administer competition law and policy are smaller when compared to larger countries.[5] Although smaller states will need a smaller number of personnel, the proportionality rule is not likely to hold, due to the problem of indivisibility (discussed further below), especially in matters associated with administration. The number of personnel and the cost of administration, per capita of population, are likely to be larger in small states when compared to larger states.

2.2.3   Market Failures and Externalities

In a small domestic market, especially in the case of islands, one is more likely to find market failures, due to a number of factors, including the existence of relatively large external social and environmental effects. In such cases, market forces cannot be relied upon to ration supply and demand. In Malta, for example, business activity tends to have relatively large environmental impacts, due to the limited land available for development. This often leads to the need to limit the number of producers, permitting existing producers to continue enjoying dominance, even if the market, small as it may be, can take more suppliers.

2.2.4   High Cost per Capita

Small states also tend to be disadvantaged due to the so-called 'indivisibility' problem, which is associated with the inability to downscale overhead costs in proportion to the size of the population. This disadvantage is related to the limited ability of small states to enjoy the benefits of economies of scale. Because of this, many small states find it impossible to compete in those types of production that carry high overhead costs, including manufacturing.

In addition, government and its institutions, being mostly overhead costs for the whole economy, are generally very costly for small states.[6] This limits the country's ability to provide adequate administrative and support structures, including regulatory institutions, to advance competition law and policy.[7]

---

[5]Brown (2010): To make matters worse, many trained specialists originating from small states often emigrate to larger countries, where their specialised services are better utilised and where remuneration is more attractive.

[6]Ibid.

[7]McKoy (2009).

2.2.5   Limited Natural Resource Endowments

Small country size often implies poor natural resource endowment and low inter-industry linkages, which result in a relatively high import content in relation to GDP.[8] In addition, there are severe limitations on import substitution possibilities.[9]

This reality often leads to domination of the market by undertakings monopolising import channels. One also often finds in small states a strong resistance by existing businesses against parallel imports and a strong lobby for exclusive dealing arrangements, on the grounds of rationalisation. In Malta, for example, resistance against parallel imports was one of the main problems relating to the introduction of competition legislation in Malta, and after its introduction, there were several complaints by firms against parallel importers.[10]

2.2.6   High Reliance on Export Markets

A small domestic market tends to give rise to high dependence on exports and therefore on economic conditions in the rest of the world.[11] A high degree of reliance on exports could be a pro-competition situation, since it implies an orientation towards free trade and the need for competitiveness in foreign markets. However, as already explained, small size renders exploitation of the advantages of economies of scale difficult, mostly due to indivisibilities and limited scope for specialisation, which give rise to high per unit costs of production. It is thus often the case that a critical size is required to enable a firm to compete in the international market, and again here, the argument for rationalisation, and against fragmentation, is a strong one.

2.2.7   Insularity and Transport Costs

Many small states are also islands, and therefore face relatively high transport costs, which are included in the price of imported industrial supplies and finished goods.

---

[8]Briguglio (1995).

[9]Worrell (1992), pp. 9–10.

[10]A case in point is when Firm A, an exclusive distributor of Macintosh computers in Malta, unsuccessfully complained to the Office of Fair Competition against another firm (Firm B). Firm B imported these computers from Sicily and sold the computers at a fraction of the price previously charged by Firm A even after adding the Sicilian wholesalers markup and the transport costs. Firm A then appealed against the decision of the office of Fair Competition in the Courts, but the decision of the Office of Fair Competition was upheld.

[11]Briguglio (1995).

Islands, being separated by sea, are constrained to use air and sea transport only for their imports and exports. Land transport is of course out of the question, and this reduces the options available for the movement of goods.

High per unit cost of transport resulting from insularity may also give rise to additional problems such as time delays and unreliability in transport services. These create risks and uncertainties in production. Such disadvantages are more intense for islands that are archipelagic and dispersed over a wide area.

Yet another problem associated with insularity is that when transport is not frequent and/or regular, enterprises in islands find it difficult to meet sudden changes in demand, unless they keep large stocks. This, in turn, leads to additional cost of production, associated with tied-up capital, rent of warehousing and wages of storekeepers.

2.3 Implications for Competition Law

The characteristics of small states just described have implications associated with competition law and policy, notably abuse of a dominant position, agreements, mergers and enforcement of the law.

2.3.1 Abuse of a Dominant Position

Generally speaking, competition legislation does not take account of economic benefits when considering abuse of a dominant position,[12] although dominance per se is not prohibited. In competition regimes modelled on Article 102 of the Treaty on the Functioning of the European Union (TFEU) abuse arising from dominance, such as limiting production, applying dissimilar conditions (including price discrimination to equivalent transactions), charging unfair prices and refusing to supply goods or services in order to eliminate a trading party from the relevant market, are generally prohibited, and once detected the undertakings responsible are sanctioned.

There could be situations where what may be considered as abuse of dominant position in a large market, need not be so in a small market particularly with regard to discrimination, excessive pricing and foreclosure of the market, as explained below.

It should be emphasised here that the arguments presented below with regard to abuse of dominant position should not be interpreted as proposing a case for allowing such abuse in small states, but to explain that maximising consumer

---

[12]However the Court of Justice of the European Union (CJEU) case law and the European Commission's decisional practice show that allegedly abusive conduct may fall outside the be defended on grounds of efficiency. See http://eur-lex.europa.eu/legal-content/EN/ALL/?uri= CELEX:52009XC0224(01), paras 28 et seq.

welfare may, in these states, require an economic analysis which takes into account the issue of small size.

2.3.1.1   Dominance and Discriminatory Conditions

Due to the small size of the domestic market, oligopolies are common in small states. In some cases letting dominant oligopolies indulge in discriminatory practices may be to the advantage of the consumer. As Gal (2001) argues, in oligopolistic markets, discriminatory pricing may work against rigid oligopolistic price structures and could result in lowering prices to the benefit of the consumers. She argues that to forbid discrimination could 'reduce efficiency and slow reactions to changed market conduct ...Discrimination in small economies, thus, merits a deeper analysis of its real effects on the market'.[13]

2.3.1.2   Dominance and Excessive Pricing

Similarly, a seemingly excessive price, when compared to the price of similar products in larger countries, may be justified in a small economy, since this may be one way in which a firm could cover costs associated with importing the product, particularly in the case of islands where transport costs tend to be relatively high, or to cover the relatively high overhead expenses associated with importing small quantities or producing on a very small scale.

The issue of transport costs is very important in this regard.[14] One implication relating to competition law and policy is that a straightforward comparison with analogous goods in nearby mainland markets may not be appropriate.

2.3.1.3   Dominance and Foreclosure of the Market

In a small state the chances of destabilisation effects of new entrants into its small domestic market is relatively high, when compared to a large state. In a small domestic market, a relatively large new entrant firm may find itself controlling a large share of the market, and this may seriously destabilise the same market. If this same firm decides to exit at short notice, possibly leaving many business creditors at a disadvantage, the business environment would be further destabilised to the detriment of consumers. It is to be expected, in such circumstances, that existing firms may tend to forestall new entrants, not only because they fear that they will lose their share of the small market, but also to reduce the chances of instability of the same market.

Although the destabilising effects of exit and entry into the market exist also in large economies, such effects are likely to more pronounced when the domestic market is small.

---

[13]Gal (2001). See also Buttigieg (1999).
[14]United Nations Conference on Trade and Development (UNCTAD) (2013).

This does not mean that barriers to entry should be encouraged, but that (a) the limited number of players that can be accommodated in a small market constrains competition possibilities and (b) the high degree of instability that arises by the entry and exit of a relatively large firm in a small market should be given due importance when assessing consumer welfare in the context of competition law.

2.3.1.4 Dominance and Refusal to Supply

Due to the constraints of replicating infrastructural facilities, there is more scope for the application of the essential facilities doctrine in small states. In a small state, a dominant firm may try to deny entry of new competitors into the market by refusing to share facilities. Competition law generally compels a dominant firm which owns a facility essential to other competitors, generally one that involves high overhead cost, to provide reasonable use of that facility. In a small state, where infrastructural facilities are costly and difficult to replicate, refusal to grant third party access to essential facilities owned and controlled by a dominant firm should be more readily checked.[15]

Thus for example, what to a German agency would not appear to be an essential facility as it could be replicated by a potential entrant who is just as efficient as the incumbent, in a small state, where sunk costs tend to be relatively high due to the indivisibility problem, the first entrant would be able to monopolise the sector.

This means that what may not constitute refusal to grant fair access in a large country could be deemed an abuse of a dominant position in a small state.

2.3.2 Implications Relating to Agreements

Competition legislation modelled on Article 101 of the TFEU, relating to agreements between undertakings, often permit restrictions in this regard, if the agreement contributes towards the objective of improving production or distribution of goods or services or promoting technical or economic progress.[16] This is the case in Maltese competition law. In other words, agreements that may appear to be anti-competitive may be exempt if, on balance, they have an overall positive impact on the consumers.

It may be argued that in a small state collaborative arrangements (horizontal as well as vertical ones) may have positive effects on the consumers, due to the advantages of business consolidation, given the very high incidence of micro-enterprise in such states.

---

[15]Buttigieg (1999).

[16]European Commission (2010): This is subject to the so-called 'pass-on requirement', meaning that consumers should ultimately get a fair share of the benefits. Furthermore, restrictions on competition must be indispensable to achieve the benefits and competition should not be substantially curtailed as a result of the agreement.

The European Commission, for instance, recognises that joint purchasing arrangements can often be pro-competitive because they allow smaller rivals to achieve similar purchasing economies to larger competitors, which can lead to enhanced competition, for example, in the form of lower prices and/or better quality products or services.[17] In general, there are two main benefits that may be considered in permitting certain types of agreements between undertakings namely (a) substantial efficiency gains (e.g. through economies of scale and scope) that are passed on to the consumer, and (b) intensification of supply competition through a better bargaining position of the firms forming the agreement.

Again here, this argument should not be construed as one that unrestricted co-operation between competitors in small states should be allowed, but that there are special circumstances, which prevail in small states, where the pass-on benefits are substantial.

2.3.3 Implications Relating to State Aid

As is well known, in general state aid is considered as a competition distortion.[18] However the TFEU makes several exceptions to this principle (refer to Art 107 (2) and (3) TFEU).

The EU General Block Exemption Regulation (GBER),[19] permits public bodies in Member States to grant state aid for a broad range of activities for relatively high outlays without these being subject to prior European Commission scrutiny, in areas of research, development and innovation (RDI), regional urban development funds, culture and heritage conservation and infrastructures for broadband, energy and sports and recreational projects. The GBER covers various categories of aid measures, including regional aid, aid for SMEs, aid for environmental protection, aid for research & development and innovation, aid for disadvantaged workers and for workers with disabilities, social aid for transport for residents of remote regions (the outermost regions plus Cyprus and Malta),and aid for sport and multifunctional recreational infrastructure. Such aid must have an incentive effect and not be granted after a project starts.

In the case of small states, especially insular ones, the case of support of these types may be stronger than in larger territories, given the high degree of market failure in small economies and the social dimension of transport in small states that are also islands. There may therefore be a case for considering state aid as permitting some form of level playing field in cases where small size and insularity have an important bearing on the cost of production.

---

[17]See Ashurst (2014) for a discussion on possible co-operation agreements between competitors under Competition Law.

[18]Article 107(1) TFEU.

[19]See http://eur-lex.europa.eu/legal-content/EN/TXT/?qid=1404295693570&uri=CELEX:32014R0651.

2.3.4   Mergers and Efficiency

In the case of mergers, Malta's Control of Concentrations Regulations state that:

> Concentrations that bring about or are likely to bring about gains in efficiency that will be greater than and will offset the effects of any prevention or lessening of competition resulting from or likely to result from the concentration shall not be prohibited if the undertakings concerned prove that such efficiency gains cannot otherwise be attained, are verifiable and likely to be passed on to consumers in the form of lower prices, or greater innovation, choice or quality of products or services.[20]

In the *Guidelines on Efficiencies* that originally accompanied Malta's Control of Concentrations Regulations,[21] it is stated that the type of efficiencies that are more likely to be cognisable and substantial than others, are efficiencies resulting from shifting production among facilities formerly owned separately, which enable the undertakings concerned to reduce the marginal cost of production as these are more likely to be susceptible to verification, concentration-specific, and substantial, and are less likely to result from anti-competitive reductions in output. Such justifications to anti-competitive behaviour are found in competition regimes in certain countries, such as the US, Canada and Australia, where the efficiencies defence is expressly mentioned in the law.

An issue relating to mergers in a small economy is efficiency consideration. In a small economy, where market dominance and natural barriers to entry are common, and sometimes cannot be easily dismantled, efficiency clauses are likely to have more significance than in larger countries. In such cases, merger control that does not sufficiently acknowledge efficiencies may actually impede restructuring of firms, in their attempt to attain a 'critical mass'.

According to the International Competition Network (ICN) many members of the ICN:

> ... made it clear that the size of the economy may ultimately affect the economic realities surrounding the merger and, in turn, the final outcome of the analysis. Numerous contributors point out that the size of the economy may also shape procedural elements of the merger control regime, such as the statutory thresholds which trigger a duty by the parties to a proposed merger to submit a pre-merger notification filing to be reviewed by the competition authority.[22]

One factor that should be considered when discussing mergers in small states relates to the notification thresholds, based on a turnover upper limit which applies to merging undertakings when they notify their merger. It makes sense that a

---

[20]Government of Malta (1993), art 4(4). See http://www.justiceservices.gov.mt/Download Document.aspx?app=lom&itemid=10475&l=1

[21]See the Control of Concentrations Regulations as first published in the Government Gazette, available at http://justiceservices.gov.mt/DownloadDocument.aspx?app=lp&itemid=18134&l=1

[22]ICN (2009), p. 31.

turnover threshold is determined in relation to the size of the economy, as otherwise if set too high, all mergers will not need notification.[23]

A third important issue in this regard relates to the benefits of networks. Such benefits acquire greater relevance in the case of sectors relating to communications and information technology. In such sectors, a concentration could enhance consumer welfare, as otherwise consumers would lose the benefit that a more extensive network could generate in such sectors, including wider choice of complementary products and enhanced quality and service that this brings about. For example, in mobile telecommunications, as more users join a particular mobile network, that network becomes more valuable to those users as they can contact more people, in more locations, at lower cost as the network expands. In the transport sector, more integrated transport services can lead to network benefits that would improve service quality through strengthened hubs, better through-ticketing arrangements, more extensive services, more comprehensive and coherent information or better co-ordination of connecting services.

The relevance of all this to small states is that the positive impact on the economy arising from mergers are likely to be more pronounced than in larger states, due to the fact that in a small market it may be desirable to avoid excessive fragmentation and encourage consolidation.

2.4   Implications Relating to the Culture of Competition

In small states, the culture of competition may not easily take root due to the fear that intense competition may destabilise a small fragile and thin market. Another reason is that, as already noted, government involvement in such states tends to loom large over the market, and public undertakings often clamour for exclusion from competition law provisions claiming that they have a social role to play. In addition, the advantages of business consolidation and the disadvantages associated with business fragmentation often lead authorities of small states to justify monopolistic and oligopolistic structures.

Furthermore, even where competition legislation is in place, in small states its enforcement may be more difficult than in larger countries due to the fact that everybody knows each other, and social and inter-family links predominate. Thus, in small states, methods other than enforcement may sometimes bring better results as far as implementing competition policy is concerned. Competition advocacy among citizens, to render them aware of the benefits of competition policy is of relevance in this regard.

---

[23]See CMS (2014) for a description of merger control regulations of 44 jurisdictions, including those in the following small states: Cyprus, Estonia, Iceland, Liechtenstein, Lithuania, Luxembourg, Macedonia, Malta, Montenegro, and Slovenia.

A World Bank/OECD (1998) report on the role of the competition agency states that such an agency '... must assume the role of competition advocate, acting proactively to bring about government policies that lower barriers to entry, promote deregulation and trade liberalization, and otherwise minimise unnecessary government intervention in the marketplace'. According to the UN Conference on Trade and Development, advocacy interventions are likely to be more effectively undertaken by the competition agency, if that agency has a stakeholder strategy, identifying the different methods and tools to be used for interaction with the different stakeholders.[24] Advocacy requires reaching the stakeholders, central and local government bodies, business representatives, consumer organisations, trade unions, professional bodies, the media, and educational establishments. In small states reaching these stakeholders may be easier to do than in larger states.

The issue of enforcement versus advocacy is a very important consideration for small states. It is not being suggested here that advocacy and enforcement contradict each other or are mutually exclusive, as in many ways they are interdependent, as argued by Clark (2004), principally because advocacy can favourably affect enforcement by fostering a competition culture, based on awareness that abuses of dominance and collusion are undesirable. Difficult as it may be in a small state, enforcement will remain important as there are always vested interests that gain from weak legal control. The argument proposed here is that advocacy, aimed principally to foster a competition culture, is of major benefit to small states as this itself encourages compliance.

2.5   Conclusion

This chapter has highlighted a number of areas which are associated with small states and which are likely to have a bearing on competition law and policy. The main argument put forward in the chapter is not that competition rules should be discarded, or that abuse should be tolerated in small states.

The basic contention is that exceptions, normally based on considerations such as improved efficiency, distribution, and overall consumer benefit, are more likely to be relevant in small states in certain circumstances.

References

Armstong HW, Read R (1998) Trade, competition and market structure in small states. Bank Valletta Rev 18:1–18
Ashurst LLP (2014) Co-operation agreements between competitors. Available at https://www.ashurst.com/doc.aspx?id_Resource=4712. Accessed 30 June 2016

---

[24]United Nations Conference on Trade and Development (2013).

Briguglio L (1995) Small island developing states and their economic vulnerabilities. World Dev 23:1615–1632

Brown DR (2010) Institutional development in small states: evidence from the Commonwealth Caribbean. Halduskultuur Adm Cult 11(1):44–65. Available at http://halduskultuur.eu/journal/index.php/HKAC/article/view/13. Accessed 30 June 2016

Buttigieg E (1999) The notion of dominance and the control of abusive pricing under Maltese competition law. Bank Valletta Rev 19:1–24

Clark J (2004) Competition advocacy: challenges for developing countries OECD and IADB. Available at http://www.oecd-ilibrary.org/governance/competition-advocacy_clp-v6-art10-en. Accessed 30 June 2016

CMS (2014) CMS guide to merger control in Europe 2014. Available at http://www.cmslegal.com/Documents/CMS_Guide_to_Merger_Control_in_Europe_2014_hi-res.pdf. Accessed 30 June 2016

Commonwealth Secretariat (2016) Small states. Available at http://thecommonwealth.org/our-work/small-states. Accessed 30 June 2016

Downes A (1988) On the statistical measurement of smallness: a principal component measure of size. Soc Econ Stud 37(3):75–96

European Commission (2010) Guidelines on the applicability of article 101 of the Treaty on the functioning of the European Union to horizontal co-operation agreements

Gal MS (2001) Size does matter: general policy prescriptions for optimal competition rules in small economies. Univ Calif Law Rev 40:1437–1478

Gal MS (2002) Reality bits (or bites): the political economy of competition policy in small economies. In: Hawk B (ed) International antitrust law and policy. Juris Publishing, New York

Government of Malta (1993) Fair trading: the next step forward. Department of Information, Malta

International Competition Network (2009) Competition law in small economies. Competition Commission Israel Antitrust Authority, Zurich

Jalan B (1982) Classification of economies by size. In: Jalan B (ed) Problems and policies in small countries. Croom Helm, London

McKoy D (2009) Competition law and policy in a small-state setting. Available at http://www.derrickmckoy.net/index.php?option=com_content&view=article&id=67:competition-law-and-policy-in-a-small-state-setting&catid=37:policy&Itemid=67. Accessed 30 June 2016

Murphy S, Smith A (1999) Competition and network utilities in small states: the example of Northern Ireland. Bank Valletta Rev 20:27–60

Muscat B (1998) Regulation of parallel behaviour in an oligopolistic market: myth or reality. Bank Valletta Rev 18:19–38

United Nations Conference on Trade and Development, The Bulgarian Commission on Protection of Competition (2013) Guidelines for implementing competition advocacy. Available at unctad.org/meetings/en/Contribution/ccpb_SCF_AdvocacyGuidelines_en.pdf. Accessed 20 June 2016

World Bank/OECD (1998) A framework for the design and implementation of competition law and policy, Chapter 6. Available at http://www.oecd.org/dataoecd/52/42/32033710.pdf. Accessed 16 Jun 2016

Worrell D (1992) Economic policies in small open economies: prospects for the Caribbean. Commonwealth Secretariat, London

Chapter 3
Small States in the UNSC and the EU: Structural Weaknesses and Ability to Influence

Baldur Thorhallsson

3.1 Introduction

This chapter enquires into the methods and tools that small states can use to influence decision-making in the European Union (EU) and the United Nations Security Council (UNSC). It examines the structural weaknesses of small states within these organisations, and how they can compensate for them in order to have a say within these organisations.

Small states are at a disadvantage relative to large states due to their lack of economic strength, administrative capacity and military capabilities, which stem from their small populations, territories and economies. Within the EU, the structural disadvantage is formalised through differential voting power in the Council whereas the structural disadvantage in the UNSC is formalised through the permanent membership of five large states (known as the Permanent Five or 'P5') and their veto powers.

While the UNSC and EU are very different organisations, small states can adopt similar strategies to exert influence in these bodies. To overcome their structural weaknesses, small states need to acknowledge their limitations, have the political will to *try* to exert influence, set priorities, develop diplomatic skills and knowledge, develop a positive image, build coalitions, take initiative and exploit the special characteristics of their small public administrations. Due to the small size of their public administrations, small states can rely on the informality, flexibility, and the autonomy of their diplomatic forces. A strategy based on these features can bring negotiating success in these organisations.

We demonstrate that while it remains useful to define influence in regional and international organisations on the basis of traditional power variables, such as large

B. Thorhallsson (✉)
Faculty of Political Science, University of Iceland, Reykjavik, Iceland
e-mail: baldurt@hi.is

© Springer International Publishing AG 2017
P. Butler, C. Morris (eds.), *Small States in a Legal World*,
The World of Small States 1, DOI 10.1007/978-3-319-39366-7_3

population, territory, economy and military strength, there are additional key factors that states of all sizes can rely on for influence and which small states are uniquely better equipped to employ.[1]

3.2   Theories of Power and Small States

The power of states has traditionally been measured on the basis of objective criteria, such as population and territorial size, gross domestic product (GDP) and military capacity.[2] In these terms, small states are held to be politically, economically and strategically vulnerable[3] and, as such, incapable of exerting any real influence in world affairs.[4]

For realist thinkers, interstate relations are guided by military strength, power politics, warfare and survival under anarchy.[5] In this pessimistic world-view, there is not much other than military capability that can restrain states and shift their reasoning. In the words of Mearsheimer, 'the fortunes of all states—great powers and smaller powers alike—are determined primarily by the decisions and actions of those with the greatest capability... great powers are determined largely on the basis of their relative military capability'.[6] Realists do not see a significant role for international norms and rules in determining state behaviour.[7] 'Institutions have minimal influence on state behavior',[8] as they are merely a reflection of 'state calculations of self-interest based primarily on the international distribution of power'.[9] Social, commercial and historical ties between states do not count for much and pose little restraint on states when interests clash.[10] Small states are therefore largely unaccounted for as international actors due to their non-existent

---

[1]Small state studies use a wide variety of definitions of state size. For the purposes of this chapter an expansive definition of small states is used in order to examine a larger pool of states to draw more robust conclusions about the challenges and opportunities facing small states. Six states are most often regarded to be part of the 'larger states' category within the European Union whereas the others are regarded as small states though they vary greatly in size based on the traditional variables defining size of states. The traditional variable, population, most often used to define the size of states is a useful tool in making a distinction between smaller and larger states within the UN. This chapter regards states with less than 11 million inhabitants as part of the smaller group category within the UN.

[2]Archer and Nugent (2002).

[3]Vital (1967).

[4]Keohane (1969).

[5]See Morgenthau (1972), Waltz (1979), Wight (1991).

[6]Mearsheimer (2001), p. 20.

[7]Schweller (2003), p. 323.

[8]Mearsheimer (1995), p. 7.

[9]Ibid, p. 13.

[10]Waltz (1959).

traditional power capabilities and the lack of motives for great powers to comply with their smaller counterparts' wishes.

Neo-liberal institutionalism provides a better and more optimistic understanding of the role that small states play in the international system. Neo-liberal institutionalists see much more room for cooperation and stability in the anarchical international system than realists do. Much of the non-cooperation is a result of collective action problems, which states can overcome. One way of doing so is through international organisations. According to Keohane and Martin, 'institutions can provide information, reduce transaction costs, make commitments more credible, establish focal points for coordination, and in general facilitate the operation of reciprocity'.[11] There are practical incentives for states, large and small alike, to cooperate. One way in which the United States (US), for instance, facilitates and maintains stability is by restraining its own power in exchange for the cooperation of other states in maintaining the international order.[12] There is consequently room for small states to have a say in world affairs, but their role is still highly circumscribed by the asymmetry of capabilities.

In the small states literature, considerable attention has been given to the more subtle and qualitative factors that matter in the exercise of power. In practice, diplomacy hinges on more than structural resources. Even small states can develop diplomatic skills and competences on certain topics, giving them issue-specific power.[13] Small states are also uniquely poised to establish positive images of themselves and good reputations, which can translate into influence.[14] The Nordic states have, for instance, been identified as norm entrepreneurs.[15] The special characteristics of the bureaucracies of small states can also be an advantage, as their diplomats are more flexible and autonomous than their counterparts from large states.[16] Diplomats from small states can therefore act quickly,[17] form relationships[18] and build coalitions.[19] There is therefore more to the exercise of power and influence than material capabilities.

---

[11]Keohane and Martin (1995), p. 42.

[12]Ikenberry (2001).

[13]Habeeb (1988), Tallberg (2008).

[14]Magnúsdóttir (2009), Ulriksen (2006), Jakobsen (2009).

[15]Ingebritsen (2006), Jakobsen (2009), Grön and Wivel (2011), Kronsell (2002).

[16]Thorhallsson (2000, 2006a), Panke (2010).

[17]Panke (2010).

[18]Thorhallsson (2006b), Grön and Wivel (2011), Katzenstein (2003).

[19]Grön and Wivel (2011).

3.3   Small States in the European Union

3.3.1   Opportunities and Constraints in the Institutional Set-Up of the European Communities

During the negotiations leading up to the creation of the European Communities, the original small member states—the Benelux states of Belgium, the Netherlands and Luxembourg—were highly concerned with their limited economic and military capabilities compared with those of the three larger members. In the end, the institutional structures took account of their demands for a more balanced decision-making system. The small states were granted a proportionally larger voice within the new institutions compared with (West) Germany, France and Italy. This arrangement, enshrined in the Treaties, mainly took the form of each member state's having a right of veto in the Council of Ministers, equal access to the policy-making structure of the European Commission and a proportionally higher number of representatives from the small countries in the European Assembly. Also, the new institutions of the Communities were mostly sited in the small countries. More generally, in the most innovative feature of the 1957 Treaty of Rome, power was transferred from the stronger (and sometimes aggressive) states to supranational bodies where the interests of small states could be better taken into account. This unprecedented institutional set-up provided the small countries with the chance to influence policy-making at the European level to an extent never seen before. It gave the small member states a flying start within the Communities.[20]

However, it soon became clear that France and Germany would retain the initiative in the process of European construction. Franco-German cooperation became the vehicle for steps towards further integration. Their role reflected their greater resources compared with the other members, and their political motivation, including how they wanted to be portrayed (their 'image') at home and abroad. The economic and administrative capabilities of the larger members, and later their military contribution to the protection of Western democracies, gave these two states and Italy a much greater say than that of their three smaller partners in the Union. The creation of the European Council, though informal in nature at first, gave heads of governments increased scope to influence the overall development of the European project, as well as the nitty-gritty details in each and every policy field. When the United Kingdom (UK), and later Spain joined, the others, claiming their own space at the negotiation table of the European Council, they in turn became more pivotal actors than the smaller newcomers Ireland, Denmark, Greece and Portugal.[21] Recently, Poland has become more assertive within the Union and has been more inclined to use the typical bargaining tactics of other large states (see below) than are the other, smaller newcomers in the EU.[22] Normally, the six states

---

[20]Thorhallsson (2004).

[21]Thorhallsson (2000).

[22]For instance, see the Government of Poland (2014).

Table 3.1 Member States of the European Union: Size Index[a]

States	Population (millions) 2014	GDP (Euro (billions)) 2013	Foreign service personnel April 2001	Year of EU entry
Small states				
Malta	0.43	7.5	256	2004
Luxembourg	0.55	45	206	1958
Cyprus	0.86	18	231	2004
Estonia	1.3	19	479	2004
Latvia	2	23	455	2004
Slovenia	2.1	36	451	2004
Lithuania	2.9	35	440	2004
Croatia	4.2	44	–	2013
Ireland	4.6	175	820	1973
Slovakia	5.4	74	931	2004
Finland	5.5	202	1642	1995
Denmark	5.6	253	1663	1973
Bulgaria	7.2	41	–	2007
Austria	8.5	323	1397	1995
Sweden	9.6	436	1500	1995
Hungary	9.9	101	1923	2004
Portugal	10.4	171	2038	1986
Czech Republic	10.5	157	2165	2004
Greece	10.9	182	1810	1981
Belgium	11.2	395	2103	1958
Netherlands	16.8	643	3050	1958
Romania	19.9	144		2007
Large states				
Poland	38.5	396	2730	2004
Spain	46.5	1049	2619	1986
Italy	60.8	1619	4688	1958
United Kingdom	64.3	2017	5500	1973
France	65.9	2144	9800	1958
Germany	80.8	2809	6515	1958

[a]Population and GDP from Eurostat. Foreign service personnel—excluding personnel employed locally by missions abroad: Information collected in foreign ministries in every member state in April 2001

mentioned so far are regarded as the EU's 'large' states, and all others as part of the small states group (see Table 3.1).[23]

---

[23]Thorhallsson (2006a).

3.3.2 The Disadvantages of Small Size

Small states face political and administrative problems within the EU decision-making processes due to their more limited resources. Caroline Grön[24] argues that a lack of resources is a key constraint even for the relatively richest and best-organised small states after joining the EU. For Jonas Tallberg,[25] small states have less 'aggregate structural power', by which he refers to economic and military capabilities, and population size. Economic size is associated with a greater influence in economic negotiations, and military capabilities with a greater influence on security policy. Population size plays a part in the perception of greater democratic representation when the EU negotiates policies.

Diana Panke[26] claims that small states face several structural disadvantages. First and foremost, they do not have 'small-state interests' but varied interests. As a consequence, small states do not have any predisposition to form coalitions and adopt joint positions, except on issues that concern size specifically (such as treaty reform). Second, small states have fewer votes in the Council. Even if the final decisions are generally taken by consensus, negotiations take place in the shadow of voting calculations, which means that larger states have a greater say in working groups, in the Committee of Permanent Representatives, in the Political and Security Committee and in the Council itself. Third, small states are less able to offer side payments than the large states due to their limited financial and economic capabilities. Fourth, a small country is seen as a less valuable coalition member than a large country. A winning coalition must have a few big states and several small states—and drawing a large state into a coalition is more valuable than accommodating a small state. Small states consequently face more difficulties in creating coalitions, and are less valuable additions to coalitions. Fifth, small states have smaller public administrations, with smaller ministries and delegations, which limits the resources that they can devote to issues and the expertise that can be formed.

3.3.3 How Small States Overcome Their Weaknesses

Small states compensate for their structural weakness in several ways. Firstly, prioritisation is the key word for small states in the EU. Small states tend to concentrate on policy sectors where direct benefits can be gained. Moreover, they focus on particular issues within these policy areas in order to guarantee their interests.[27] For instance, Luxembourg has put great effort over the years into

---

[24]Grön (2014).

[25]Tallberg (2008).

[26]Panke (2010).

[27]Thorhallsson (2000), Panke (2010).

securing favourable deals for its financial sector. The Baltic states have been giving priority to secure energy supplies from the West, the smooth adoption of the euro and whatever degree of protection they can gain from the EU's defence and security policies. Cyprus is an example of a small state that uses its strength mainly in connection with resistance against favourable EU policies towards Turkey and occupied North Cyprus.

Small states are forced to prioritise because of the small size of their administrations. They cannot take part in all the EU's activities, and must reluctantly admit that they have to miss some meetings within the EU institutions due to their limited staff numbers. This is particularly the case with preparatory panels and comitology committees in the Commission.[28] Moreover, officials from small states often attend meetings simply to observe the ongoing debate without any intention of influencing it. To counter this problem, Luxembourg has an arrangement with Belgium whereby it can be represented by the latter in several meetings.[29] Also, Luxembourg has required practical assistance from the Netherlands during its tenures of the EU Council Presidency; for reasons of convenience, the presidencies of Luxembourg and the Netherlands are always paired in sequence with each other.[30]

On the other hand, the narrower range of small states' interests within the EU makes it not only viable but also practical for them to prioritise more strictly. The economies of the small member states rely on fewer export products. Two or three agricultural products under the Common Agricultural Policy may, for instance, account for a much larger share of their total agricultural production than of that of the large states.[31] Hence, small states can prioritise to a much greater extent than the large ones, without damaging their interests.

Secondly, within small states, the working procedures for handling EU affairs are characterised by informality and flexibility. The small size of the bureaucracy allows for smooth and efficient decision-making. Officials tend to know and trust each other. Decisions are often taken in informal meetings or over the telephone. On the other hand, officials in the EU delegations of small states claim that they often lack information and clear objectives from their capitals. Normally, they receive guidelines instead of instructions, which may also be oral rather than written. They are formally and informally expected to find their own way of participating in the EU decision-making process. Hence, officials from small states have greater autonomy and are expected to take their own initiatives in order to succeed in EU negotiations. Still, if needed, they can easily contact higher-ranking officials who have the authority to alter major decisions. Moreover, they will receive written instructions from their ministries if the issue at stake is of great importance for the country—as is common practice in the case of larger states.

---

[28]Lægreid (2000).

[29]Thorhallsson (2000).

[30]Van den Berg (1994), Thorhallsson (2006b).

[31]Thorhallsson (2004).

Officials in small states' permanent representations to the EU, located in Brussels, play a greater role than do their counterparts from large states. They participate in domestic policy-making in their capitals as well as in Brussels. They are more often responsible in their own right for negotiations in the Council than is the case with diplomats from the large states, which more often send experts from ministries to negotiate on their behalf. An official from a small state is expected to specialise in a particular issue or policy field—but is also required to have a good overview of EU policies and attend meetings on a range of topics, as will be discussed below.

Officials in the public administrations of small states dealing with EU affairs tend to have much broader responsibilities than do their counterparts from the large states. The same official in a small country may participate in domestic policy-making while also contributing to policy formation in the Commission. He or she may negotiate in Brussels on behalf of his or her state and take the final decision on a given Commission proposal in one of the working groups of the Council. The same official will sometimes advise the minister in the Council and have responsibility, with others, for deciding on criteria and guidelines in the comitology committees run by the Commission. Finally, the same official may advise on or take a direct part in the implementation process of the resultant directive in his or her home capital. This is never the case with the six large member states. Their larger bureaucracies operate with a much clearer division between policy-making, negotiating and the implementation process. Interestingly, in the mid-1990s, the Netherlands—which had not required their EU staff to handle the same variety of functions across these divisions and phases of the Commission's work as had the other small states—made explicit efforts to adopt this small-state tactic in order to cope with the growing EU workload.[32]

Thirdly, the European Studies literature has generally held that coalition building has been crucial when small states have succeeded in influencing EU policy.[33] Panke[34] argues that several features can help small member states to punch above their weight within the EU: these include years of membership (lengthier membership translates into greater knowledge), domestic political stability, priority-setting (as discussed above), the use of a broad range of negotiation strategies, clear negotiating instructions (a clear national position), and active participation in negotiations. Small states have especially good chances of succeeding in negotiations, according to Panke, if they use persuasion-based strategies rather than bargaining-based ones; can demonstrate a decisive level of knowledge; and present their case as entailing benefits for other member states and for the greater European good.

Small states in general have two possible negotiating strategies. One, they can become proactive in EU policy-making only when issues of direct national interest

---

[32]Thorhallsson (2000).

[33]Thorhallsson and Wivel (2006), Arter (2000), Maes and Verdun (2005), Wivel (2005), Duke (2001).

[34]Panke (2010).

are on the agenda. They are concerned not to become isolated and so are willing to compromise. However, and secondly, they will be reactive and flexible regarding issues of little national interest to them. Here they will devote all their administrative capacity to guaranteeing a favourable outcome in sectors of importance.[35]

The larger states tend to adopt an inflexible negotiating strategy on all occasions. Their veto threat is taken seriously—unlike that of smaller states. Consequently, the small states tend to emphasise the culture of consensus, and prefer bargaining in the form of 'package deals' where they can give priority to a few issues of importance while setting others aside. By contrast, the larger states have sufficient administrative capacity to focus on all the EU's policy sectors and have a wider range of interests within it, including as regards controlling EU expenditure and securing their own international positions. Networking and coalition building are central to all member states, but small states have a greater need to form alliances with other members and the Commission (as discussed below) due to their lesser capacity and reduced political clout, also as regards veto power.[36]

Fourthly, small states will use the features identified above also when working with and within the European Commission. Their *relations with the Commission* are characterised by their reliance on it, whether working in its own committees or in the Council. This is because small states often lack information in policy areas of limited domestic importance, and are in greater need of Commission knowhow and guidance in negotiations in the Council than are the large states. As a result, small states tend to cooperate with the Commission whereas the large states are much more confrontational towards it.[37] At the same time, small states face various constraints when lobbying the Commission due to their limited resources and lack of focus on Commission processes: this has been shown at least in the case of two small states, Denmark and Sweden.[38]

Small states try to influence proposals in the initial stages within the Commission's committee system.[39] They will call upon the Commission to act as a mediator between them and the large states, and generally prefer to deal with issues at the Commission level rather than having to negotiate with all EU member states in the Council.

The special features of small public administrations, such as informality and flexibility, help the small states to develop 'a routine working process' with the Commission. A few officials, often just one or two, in a small country will be in direct contact with Commission officials regarding any given proposal. On the Commission side, there will often be only one rapporteur dealing with the particular proposal. Hence, a close working relationship can be established between these actors on each side of the table, and make communication and decision-making

---

[35]Thorhallsson (2000).

[36]Ibid.

[37]Ibid.

[38]Grön (2014).

[39]Thorhallsson (2000), Panke (2010).

much smoother than in the case of the large states. In the process, the Commission can gain special insights into the situation in these small countries. It is particularly relevant for the Commission to develop good relations with the officials of small states, precisely because they tend to be personally involved at all levels of EU affairs. These officials can make an important contribution to the drafting of proposals and can often respond swiftly to new developments in negotiations, making decisions autonomously. Such relations can facilitate mutual understanding between the small state and the Commission, giving the former better chances of getting its views incorporated in the Commission's policy proposals. It is true that a larger state may be a more valuable coalition partner for the Commission—but it is often much cheaper 'to buy' the support of a small state than a big one. That said, the Commission is not ultimately or in any simple sense the defender of small-state interests. Also, small states must hold their own against the Commission if they want to convince it that their ideas should be incorporated into its proposals.[40]

Findings by Grön indicate that civil servants from two small states, Denmark and Sweden, regard technical propositions as the most efficient way to convey their inputs, along with proposals that embody the general good of the EU and can lead to a compromise. Grön argues that Denmark and Sweden can impact legislation considerably if they lobby the Commission early on in the decision-making process and provide technical input.[41]

In the new College of Commissioners appointed in 2014, small states are in a particularly strong position. The President of the Commission comes from one of the smallest member states (Luxembourg) and six out of its seven Vice-Presidents are also from small states (Finland, Latvia, Slovakia, Estonia, Netherlands and Bulgaria). The sole exception is the High Representative of the Union for Foreign Affairs and Security Policy, who comes from Italy. The media and even the academic world would be likely to conclude that the large states had taken over the Commission if it had been they who filled most of these posts.

Fifthly, small states build their international image on their domestic reputation and knowledge in particular policy fields.[42] Ingebritsen[43] argues that small states can become norm-setters in regional and international organisations. For instance, domestic success and deep-rooted knowledge on certain internationally salient issues such as environmental protection, human rights, women's rights, and humanitarian and development assistance have had enormous value for the Nordic states' position within the EU. These states have created a positive, progressive and responsible image of themselves internationally, using it to place issues on the agenda, build coalitions, and get their ideas accepted within the EU. They have thereby become what the literature calls 'norm entrepreneurs'. For instance, the three Nordic states have managed to use their favourable image as environmentally

---

[40]Thorhallsson (2000).

[41]Grön (2014).

[42]Thorhallsson (2006b).

[43]Ingebritsen (2006).

friendly countries to make their voice heard, and even to take the lead, within EU environmental policy.[44]

Furthermore, a stable political system and good governance at home can become a soft-power tool. This has often held good for the Benelux and Nordic member states within the EU, who regularly rank near the top of the Vision of Humanity organisation's Global Peace Index, which measures both internal and external behaviour. For similar reasons, personalities from small states are often entrusted with mediating or investigatory duties that raise both their own and their nation's profile: a practice also reflected in the EU's choice of Special Representatives, other emissaries, and commanders for Common Security and Defence Policy missions.

However, small states need to have a certain critical mass of administrative competence to transform their positive image in a specific policy field into a power resource.[45] A focus on making technical inputs on prioritised proposals can be particularly cost-effective for them. Small states can punch above their weight if they are able to use their knowledge as a power tool in providing helpful information for other members and the Commission, thereby gaining especially good chances of influencing the technical side of proposals.[46]

Finally, states also need to have the political will and ambition to take an active part in EU decision-making. Some states may simply lack ambition in this respect,[47] as has been seen (for different reasons) in the cases of Greece[48] and Slovenia.[49] Political leaders need to be willing to spend time, effort and money on working within the EU.[50] Moreover, the governing elite must believe that it can have a say in the decision-making processes. If it considers its chances hopeless and/or sees the EU as doomed to be ineffective on a given issue, it may not even try to contribute to or influence decision-making.[51] But there are positive examples: most members of the Danish political elite appear to be confident of guaranteeing Danish interests within the EU; they want to keep Denmark at the heart of the European integration process.[52] Also, the rotating presidency of the Council provides small states with a good opportunity to gain influence on 'low politics' within the Union by acting as a mediator. However, small states need to have the political will and ambition to use the Presidency as a power resource. On the other hand, small states still have difficulties influencing 'high politics' issues through the new Council President and the High Representative for Foreign Affairs and Security Policy. These issues continue to be dominated by large states.[53]

---

[44]Magnúsdóttir (2009).

[45]Thorhallsson (2012, 2013).

[46]Grön (2014).

[47]Thorhallsson (2006b).

[48]Thorhallsson (2000).

[49]Toplak (2014).

[50]Grön (2014).

[51]Thorhallsson (2006b).

[52]Petersen (1998).

[53]Grön and Wivel (2011), p. 533; Tallberg (2008), p. 697.

Personal leadership skills and ability to take initiatives are of enormous value in such cases when combined with political willingness to influence EU decision-making. This has been clearly demonstrated, as noted, with regard to the impact of the three Nordic EU member states on environmental policy. The ECFR scorecard for foreign policy activism furthermore shows that states with small populations need not necessarily lag behind states with large populations when it comes to foreign policy activism.[54]

3.4   Small States in the United Nations Security Council

3.4.1   *Opportunities and Constraints in the Institutional Set-Up of the United Nations*

The founding of the UN reflects a compromise between the great powers at the end of WWII and other states. The compromise was, however, highly favourable for the great powers, as the inequality of states is formalised in the UN Charter. The UN Security Council is the most powerful body of the UN and 'potentially the most powerful international organization ever known to the world of states'.[55] The original UN Charter gave the victors of WWII (the United States, Soviet Union, United Kingdom, France, and the Republic of China) a privileged position in the UNSC: permanent membership and veto rights.[56] The other six members of the UNSC were elected on two-year terms by the General Assembly.[57] The asymmetry in voting and seating arrangements in the Security Council came about through the desire of the great powers at the end of WWII to entrench their dominance.[58] Other states accepted this, as it was understood that there would either be a UN with great-power privilege or no UN at all.[59] Despite numerous reform attempts, the voting and seating arrangements of the UNSC have been remarkably stable, changing only in 1965 when the non-permanent members were increased to ten.

While the UNSC formalises the inequality of states, there are still institutional constraints on the dominance of the P5 and room for other states to exercise influence. The first constraint is disunity among the P5. The Cold War is a good illustration of such deadlock. As a result of the Cold War, the Security Council was incapable of playing a role in conflicts unless they were unrelated to the interests of the great powers. When the P5 are united, the second constraint is the so-called

---

[54]European Council on Foreign Relations (2016).

[55]Hurd (2007), p. 12.

[56]United Nations Charter 1945, art 23.

[57]Ibid.

[58]Krisch (2008), pp. 135–136.

[59]Ibid, p. 136.

'sixth veto', referring to the ability of the non-permanent members to prevent resolutions from being passed. As any resolution needs nine affirmative votes, non-permanent members need to be persuaded. The increase in non-permanent members and the change from a seven-vote threshold to a nine-vote threshold has consequently strengthened the ability to exercise the sixth veto, which explains why some of the P5 resisted changes and tried to limit the increase in members and the threshold for affirmative votes.[60] Also, the veto power of the P5 does not apply to procedural matters, which means that any member can introduce a matter in the Security Council, forcing the Security Council members to take a position on the matter. For Hurd, this represents a significant source of power.[61]

Besides the UNSC, the set-up of the UN also enshrined another form of inequality, as some states were precluded membership to the organisation on the basis of their small size. These states were not granted membership because they were understood to be unable to fulfil the UN's Charter criteria due to their limited resources. The opponents of small state membership cited Article 4 of the Charter, stating that only states, which in the judgement of the Organisation, are able and willing to carry out its obligations, such as to take effective collective measures for the prevention and removal of threats to the peace, could join the UN. It was not until the early 1990s when microstates, such as Liechtenstein, San Marino and Monaco, were finally granted full membership of the UN. Figure 3.1 shows the 193 UN member states categorised according to population in millions.

3.4.2 The Disadvantages of Small Size

From 1991 to 2017, 37 countries with fewer than 11 million inhabitants (in 2014) were elected to participate in the Security Council, as Fig. 3.2 and Table 3.2 show. Of these, three had a population of between 500,000 and 1 million (Cape Verde, Djibouti and Qatar), 21 had a population between 1 and 5 million and 13 had a population between 5 and 11 million (see Fig. 3.2). It is noticeable that none of the 27 states with a population of less than 500,000 has been elected to the UNSC. Table 3.2 shows all the states with fewer than 11 million inhabitants which were elected to the UNSC from 1991 to 2017.

---

[60]Bourantonis (2005), pp. 61–62. However, there are not good reasons to expect that the vote of any single non-permanent member will matter much. Situations where a single vote either breaks or makes a resolution are scarce [Vreeland and Dreher (2014), pp. 7–8]. But the increase in non-permanent members can matter when it comes to unanimous UNSC resolutions. Since UNSC resolutions are desired for reasons of legitimacy [Claude (1966); Voeten (2005); Hurd (2007)], resolutions carry more weight when there is unanimous support for them [Krisch (2008), p. 139]. This may give non-permanent members the ability to put an imprint on proceedings.

[61]Hurd (2002), p. 40.

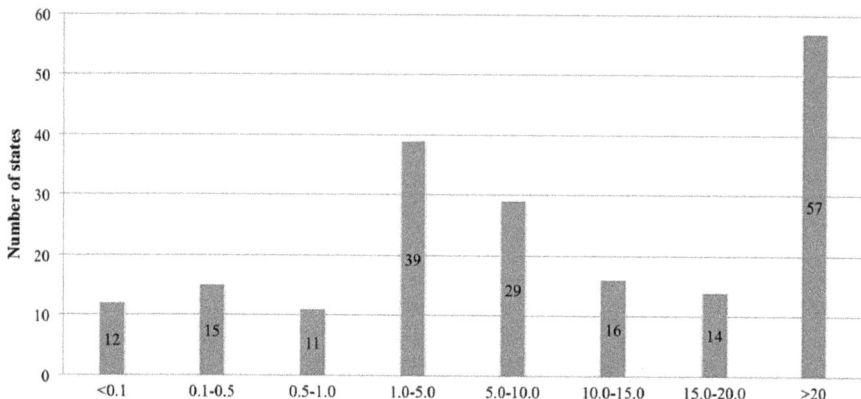

Fig. 3.1 Number of UN states (193) according to population in millions. Based on World Bank population figures for 2014

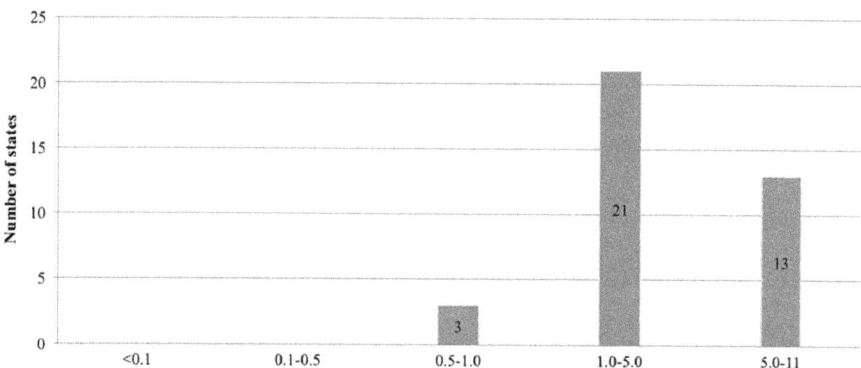

Fig. 3.2 Number of states with fewer than 11 million inhabitants elected to the UNSC, 1991–2017. The population data is based on the year 2014 (World Bank), which means that most states had smaller populations when they held non-permanent memberships of the UNSC

Small states are effectively at a disadvantage as regards being elected as non-permanent members. Findings indicate that large population size, contributions to peacekeeping, economic strength, foreign aid provision and political affiliation are associated with non-permanent membership. The only overruling factor is the norm of 'turn-taking' which gives preference to states that have never held a seat on the UNSC.[62]

As in the EU, small states' lack of aggregate structural power places them at a disadvantage in influencing affairs in the UNSC. Side payments are a prominent feature of the UNSC. It has, for instance, been shown that non-permanent members can expect more favourable decisions from the World Bank[63] and the International

---

[62]Dreher et al. (2013).

[63]Dreher et al. (2009a).

Table 3.2 States with fewer than 11 million inhabitants elected to the UNSC, 1991–2017[a]

States	Population (2014)	Period	No. of countries' UN mission staff (2010)
Sweden	9.7	1997–1998	42
Portugal	10.4	1997–1998/ 2011–2012	39
Norway	5.1	2001–2002	32
Denmark	5.6	2005–2006	21
Hungary	9.9	1992–1993	18
Libya	6.3	2008–2009	17
Slovakia	5.4	2006–2007	16
New Zealand	4.5	1993–1994/ 2015–2016	15
Czech Republic	10.5	1994–1995	14
Benin	10.6	2004–2005	11
Slovenia	2.0	1998–1999	10
Panama	3.9	2007–2008	10
Bosnia Herzegovina	3.8	2010–2011	10
Ireland	4.6	2001–2002	10
Lebanon	4.5	2010–2011	9
Costa Rica	4.8	1997–1999/ 2008–2009	9
Croatia	4.2	2008–2009	9
Botswana	2.2	1995–1996	8
Singapore	5.5	2001–2002	8
Djibouti	0.9	1993–1994	7
Oman	4.2	1994–1995	7
Cape Verde	0.5	1992–1993	7
Gabon	1.7	1998–1999/ 2010–2011	–
Qatar	2.2	2006–2007	6
Bahrain	1.4	1998–1999	4
Mauritius	1.3	2001–2002	4
Jamaica	2.7	2000–2001	2
Guinea Bissau	1.8	1996–1997	–
Gambia	1.9	1998–1999	–
Namibia	2.4	1999–2000	–
Congo Brazzaville	4.5	2006–2007	–
Bulgaria	7.2	2002–2003	–
Honduras	8.0	1995–1996	–
Tongo	7.1	2012–2013	–
Azerbaijan	9.5	2012–2013	–
Jordan	6.6	2014–2015	–

(continued)

Table 3.2 (continued)

States	Population (2014)	Period	No. of countries' UN mission staff (2010)
Luxembourg	0.6	2013–2014	–
Lithuania	2.9	2014–2015	–
Uruguay	3.4	2016–2017	–
Austria	8.5	1991–1992/ 2009–2010	–

[a]Population figures from the World Bank. When empty, the figures on UN staff were not available

Monetary Fund,[64] higher US economic aid,[65] greater German aid[66] and larger loans from the Asian Development Bank,[67] which suggests that vote trading occurs and that economic might counts for influence.[68]

Smaller states are also to a large degree ignored during negotiations, as most UNSC decisions are essentially made through the 'unrecorded and informal consultations between subsets of the permanent members'.[69] While the increase in non-permanent members increased the representation of smaller states and gave them a voice in the most important UN body, their real influence on UNSC resolutions may have declined. With a greater number of non-permanent members and a slight increase in the voting threshold, it has become easier for the P5 to find non-permanent members to support resolutions. The increase in non-permanent members consequently gives each non-permanent member a lesser ability to block measures, and makes it harder for non-permanent members to coordinate.

3.4.3   How Small States Overcome Their Weaknesses

Small member states of the UN Security Council can compensate for their weakness in several ways. Firstly, a small state has to have the political will to play an active role in the UN's work, set an aim of involvement and prioritise the UN cause. Several factors may contribute to the desire to be active in the UN. Values and ideology in the Nordic states, for instance, provide an ideological basis for their participation in humanitarian and peacekeeping missions. In addition, the world-

---

[64]Dreher et al. (2009b).

[65]Kuziemko and Werker (2006).

[66]Dreher et al. (2013).

[67]Lim and Vreeland (2013).

[68]Research suggests that vote-buying also occurs in the UN General Assembly. Studies show that voting in the UNGA is influenced by Chinese trade [Flores-Macias and Kreps (2013)], IMF and WB programs and loans [Dreher et al. (2013)] and US aid [Dreher et al. (2008)].

[69]Voeten (2005), p. 537.

views of foreign policy elites and politicians matter, as they need to believe that their state is capable of making a difference within the UN apparatus. The cohesion and national unity evident in the Nordic states' foreign policy objectives (in the case of Norway, Sweden, Denmark and Finland) towards the UN contributed greatly to their commitment to the UN cause.[70]

The Nordic commitment is illustrated by the extensive financial support for UN organisations such as the UN Office for the Coordination of Humanitarian Affairs (OCHA) and the UN High Commissioner for Refugees (UNHCR). Norway and Sweden were the second and fourth largest donors to OCHA in 2004, with contributions of 12 and 9 million dollars respectively, while Denmark and Finland were ranked as number 12 and 14, with contributions between one and two million dollars—far more than Japan and Germany. In general, the Nordic states have an astonishing track record of supporting the UN, providing, for example, 25% of all the military personnel deployed in UN peacekeeping operations during the Cold War.[71] In addition, Norway, Sweden and Denmark have, for a number of years, been among the few countries, along with Luxembourg and the Netherlands, to reach the UN target of 0.7% of GDP in Overseas Development Aid (ODA). The high ODA score and its associated 'good international practice' is an asset quite effectively utilised by the Nordic states on the international scene.[72]

A government which does not regard the UN apparatus as a means of influence is unlikely to channel resources towards the UN. This was the case with the Icelandic government, which was divided on its bid for UNSC membership. Its Prime Minister (1991–2004), who later became Foreign Minister (2004–2005), argued that Iceland, as a small state, would not be able to exercise any influence within the Security Council or derive any benefits from membership. Hence, he and some other Conservative MPs saw no reason to continue the campaign and considered withdrawing from it. When they failed to get the government to withdraw from the bid they limited the resources for it. Iceland's challengers, Turkey and Austria, openly questioned the country's seriousness about the application, which seriously undermined the country's campaign.[73]

The Norwegian Ministry for Foreign Affairs directly links reputation to influence, and the Swedish Ministry for Foreign Affairs has bluntly stated that 'Creating a strong image of Sweden abroad is another means of promoting Sweden'.[74] Thus, in the UN, the Nordic states try to influence by active participation and contributing towards humanitarianism, world peace and burden sharing. Ingebritsen argues that the Nordic states pursue 'social power' by acting to promote a particular view of the good society.[75] Hence, they act as norm entrepreneurs, persuading states to adopt

---

[70]Thorhallsson (2006b).

[71]Danish Institute for International Studies (2005).

[72]Ulriksen (2006), p. 14.

[73]Utanríkisráðuneytið (2009).

[74]Cited in Ulriksen (2006), p. 10.

[75]Ingebritsen (2006).

new norms. They are successful if states conform to their norms in the absence of domestic pressure. Nordic strategies are grounded in a commitment to pursue interests through 'soft power'. For Nye, soft power is influence derived from a state's diplomatic skills, culture and reputation as opposed to realist military and political hard power.[76] It is the ability to get others to want what you want through appeal and attraction, as opposed to coercion.

The high quality and long experience of the Nordic countries in the UN's work are also represented in UN leadership positions and within the UN bureaucracy. Finland's former president, Martti Ahtisaari, assumed the lead in the effort to determine the future status of Kosovo—as Kofi Annan's special envoy in 2005—and served as a mediator in Aceh, Indonesia.[77] Kai Aage Eide, a Norwegian diplomat, was appointed the UN Special representative to Afghanistan and Head of the UN Assistance Mission in Afghanistan from 2008 to 2010,[78] only to be succeeded by Swedish diplomat Staffan de Mistura in March 2010.[79] Other well-known Nordic persons involved in the UN's work include Hans Blix, of the United Nations Monitoring, Verification and Inspection Commission, and Jan Egeland, the United Nations Undersecretary-General for Humanitarian Affairs and Emergency Relief Coordinator. The Nordic countries are also among the most over-represented states in the UN Secretariat staff.[80]

Secondly, the importance of small states' diplomatic skills and knowledge in UN work is demonstrated by the number of times countries have been invited as outsiders to the UNSC meetings on the basis of Rule 37, as shown in Table 3.3. According to this rule, states which either have a particular interest in issues being discussed in the UNSC or have brought the matter to the Council's attention may be invited to, or request an invitation to, the UNSC without the right to vote.[81] The ability of non-members to exert influence on specific issues has increased with the publishing of the Council's monthly programme of work and a provisional agenda in advance of its meetings,[82] making it possible for states to be informed and prepare for proceedings. Importantly, countries which score high on the Human Development Index (HDI) are more likely to be proactive within the UN (see Table 3.3). A state's high ranking on the HDI is seen as a significant source of prestige and enhances its international image (this is demonstrated in newspaper coverage around the globe, for example the Nordic states are often hailed for their high HDI ranking).

Norway was invited 78 times to the UNSC in 2000–2007, despite not being part of any conflict, as Table 3.3 shows. Only Israel was invited more often. However,

---

[76]Nye (1990).

[77]Whitfield (2007), pp. 35, 268.

[78]United Nations (2008b).

[79]United Nations (2010b).

[80]Novosad and Werker (2014).

[81]United Nations (1983).

[82]Hulton (2004), p. 245.

Table 3.3 States with fewer than 20 million inhabitants invited 25 times or more to the UNSC under rule 37, 2000–2007[a]

States	Population (2010)	HDI (ranking)	Number of invitations 2000–2003	Number of invitations 2004–2007	Total number of invitations 2000–2007
Israel	7.4	0.932 (23)	33	47	80
Norway	4.7	0.968 (2)	26	52	78
New Zealand	4.3	0.943 (19)	44	30	74
Lebanon	4.1	0.772 (88)	14	53	67
Portugal	10.7	0.897 (29)	39	22	61
Cuba	11.5	0.838 (51)	29	31	60
Rwanda	11	0.452 (161)	30	22	52
Sierra Leone	5.2	0.336 (177)	22	25	47
Liechtenstein (2009)	0.035	0.951 (19)	16	31	47
Austria	8.2	0.948 (15)	15	28	43
Bosnia and Herzegovina	4.6	0.803 (66)	17	23	40
Switzerland	7.6	0.955 (7)	11	28	39
Jordan	6.4	0.773 (86)	25	12	37
Serbia	8	0.826 (67)	7	30	37
Sweden	9.1	0.956 (6)	29	8	37
Albania	3	0.801 (68)	19	17	36
Burundi	9.8	0.413 (167)	19	24	35
Netherlands	16.8	0.953 (9)	7	28	35
Finland	5.2	0.952 (11)	6	28	34
Croatia	4.5	0.850 (47)	22	12	34
Libya	6.4	0.818 (56)	25	9	34
Belgium	10.4	0.946 (17)	31	3	34
Tunisia	10.6	0.766 (91)	19	10	29

(continued)

Table 3.3 (continued)

States	Population (2010)	HDI (ranking)	Number of invitations 2000–2003	Number of invitations 2004–2007	Total number of invitations 2000–2007
Costa Rica	**4.5**	**0.846 (48)**	**17**	**12**	**29**
Chile	**16.7**	**0.708 (40)**	**19**	**10**	**29**
Somalia	*10.1*	*–*	*4*	*24*	*28*
Guatemala	*13.6*	*0.689 (118)*	*3*	*25*	*28*
Singapore	**4.7**	**0.922 (25)**	**11**	**17**	**28**
Namibia	*2.1*	*0.650 (125)*	*16*	*9*	*27*
Liberia	*3.7*	*–*	*6*	*21*	*27*
Georgia	*4.6*	*0.754 (96)*	*11*	*16*	*27*
Ireland	**4.6**	**0.959 (5)**	**4**	**23**	**27**
Senegal	*12.3*	*0.342 (156)*	*16*	*10*	*26*
Bahrain	*0.7*	*0.866 (41)*	*21*	*5*	*26*
Denmark	**5.5**	**0.949 (14)**	**22**	**3**	**25**

[a]Figures from HDR (2007/2008), HDR (2009), United States Census Bureau (2010), UNSC (2003, 2007)

Norway's total number of invitations might actually have been higher had it not been an elected member of the Council in 2001–2002. During the same period, Bangladesh, which is the largest contributor of troops to UN peacekeeping missions with nearly 11,000 police, experts, and troops in 2010 and a population of 156 million, was only invited 30 times to the Council.[83]

Of the larger states, Japan was invited a total of 155 times over the same period, India 79 and Germany 61.[84] Comparing the small states in Table 3.3, it is possible to detect a trend. Those countries, such as Israel, which have a stake in the proceedings by being conflict parties, neighbouring states or prone to internal conflicts, are invited often (distinguished by italic rows). Countries which are not party to conflicts and enjoy relatively high levels of development, such as Norway, New Zealand, Portugal, Liechtenstein, Netherlands, Finland, Belgium, Ireland and Denmark (distinguished by bold rows), are invited equally many times, suggesting

---

[83]United Nations (2010a).
[84]United Nations Security Council (2003, 2007).

that these states' rates of participation are related to particular knowledge and expertise in certain policy fields.

Thirdly, *image* and *reputation* affects what kind of influence small states can have. Small states, particularly those without a colonial legacy, are perceived to be more neutral and consequently more acceptable as mediators. Perceived historical neutrality has benefitted the Nordic states in their international work, in particular in Africa, distinguishing them from other active small states, such as the Netherlands, Portugal and Belgium, whose intentions will always be tainted by the colonial legacy.[85] However, a state's perceived neutrality is not automatically transformed into an asset for international influence. In order to become an asset, it must be combined with experience and skills, which take time, leadership and initiative to develop. Norway's reputation and engagement with the UN's conflict resolution work expanded with the Oslo process in the Middle East in the early 1990s. Since then, Norway's experience includes facilitating peace talks in the Sri Lankan civil war and Philippines civil conflict. According to Whitfield, Switzerland, the Netherlands and Sweden have followed in Norway's footsteps by also establishing themselves as international peacemakers.[86] Each of these small states frequently attends UNSC meetings under rule 37, as Table 3.3 shows.

Norway's ability to assume such a prominent position in conflict resolution is due to its long-term dedication to such causes and the perception of the country as an international norm setter. In fact, all the Nordic states are seen as norm entrepreneurs in the fields of human rights, development assistance, women's rights, participation in peace operations and humanitarian efforts and environmental protection.[87] This perception is related to the Nordic states' long history of military, police and civilian support for UN peacekeeping operations such as ONUC (1960–1964), UNAVEM I-III (1988–1999), UNTAG (1989–1990), ONUMOZ (1992–1994), UNSOM I-II (1992–1995) and UNOMSIL (1998–1999), to mention but a few.[88] More recently, Norway, Sweden, Denmark and Finland, in collaboration with Austria, Canada, the Netherlands and Poland, established the UN Standby High Readiness Brigade (SHIRBRIG) with headquarters situated in Copenhagen,[89] in effect bringing UN operations and decisions closer to home. This indicates the importance of the Nordic states within the UN system whilst strengthening the Nordic position and image.

As already mentioned, this image has been of enormous value for the Nordic countries in international affairs,[90] and they have used it actively to build coalitions and place issues on the UNSC agenda.[91] The Nordic countries have, over the years,

---

[85]Ulriksen (2006), pp. 11–12.

[86]Whitfield (2007), p. 42.

[87]Ingebritsen (2006).

[88]Ulriksen (2006).

[89]Ulriksen (2006), p. 14.

[90]Ulriksen (2006).

[91]Hansson (2007).

jointly initiated a number of UN reform projects. The Nordic UN Project's aim was to transform the United Nations Development Programme from a funding mechanism into a stronger developmental institution.[92] The latest reform attempt can be found in the high level panel entitled Deliver as One, from 2006, which tried to provide some critical answers as to why the UN has become fragmented and weak.[93]

Fourthly, prioritisation is of key importance. Small states, with less administrative capacity for collecting and analysing information,[94] have to rely on the UN Secretariat and other UNSC members to inform their decision-making. Thus, knowledge—with its preparation and prioritisation—remains one of the most important and demanding aspects of UNSC membership. It can be seen from Table 3.3 that it is not necessarily economic growth or military strength per se which determines attendance of the UNSC meetings, but the human factors behind knowledge, expertise and diplomatic skills. Thus, acquiring and building the necessary knowledge is important for a small state to become an active and influential participant as a member of the UNSC. To deal with the structural disadvantage and to maximise influence, small states have to prioritise, delegate and decide upon which issues are manageable, and which can be dealt with more effectively by others in the Council.[95]

Ireland's period in the Council was considered a success due its pragmatic approach to prioritising workloads. Ireland knew that it could not change the procedural functioning of UNSC operations and, accordingly, prioritised those tasks which lay in areas over which it believed it could exert an influence.[96] One of the successes included its robust stance against the proposition to lift the arms embargo when combatants in the Ethiopia/Eritrea conflict had reached an agreement, this position eventually winning the support of the other member states.[97]

Success is also highly dependent on a state's preparation in building the necessary knowledge base for its foreign service to be able to deal with the demands of the Council. Sweden's preparatory work included the construction of a database of the issues on the agenda corresponding to the position of different members. Subsequently, the knowledge compiled in the database was used to construct mini-seminars for the Swedish delegation leading up to its term on the Council.[98] Although unsuccessful, by 2008, Iceland's preparation for UNSC membership comprised the making of a catalogue of issues and analyses of principal subjects of discussion in the Security Council.[99] Another successful strategy is evident in

---

[92]Odén (2010), p. 273.

[93]Odén (2010), pp. 273–27 citing the Nordic United Nations Project.

[94]Gylfason (2004).

[95]United Nations Security Council (2004).

[96]Gillissen (2006), p. 37.

[97]Ryan cited in Gillissen (2006), p. 34.

[98]Rydberg (1998).

[99]Gísladóttir (2008).

Norway's preparation for membership, which included close cooperation with its knowledge institutions, such as universities, research institutes and non-governmental organisations.[100] For instance, despite Liechtenstein's being one of the smallest UN members in terms of population, it has built up a good reputation for knowledge and expertise through initiatives such as the Princeton University-based Liechtenstein Institute on Self-Determination—with direct links to the Permanent Mission of Liechtenstein at the UN. The long-serving officials at the Mission have built up capacity to take an active part in discussion on issues such as 'Civilians in armed Conflict',[101] 'Women, Peace and Security'[102] and 'Post-conflict Peace building'.[103] As Table 3.3 shows, Liechtenstein has participated many times in the Security Council proceedings under Rule 37. These cases highlight a general feature of UNSC membership: states have to commit to serious competence building and training of officials to respond to possible scenarios. In addition, to be able to handle excess workloads, states need to increase staff capacity.[104] The high number of Belgian UN mission staff, compared with other small states, is a result of its being elected to UNSC membership twice since 1991 (see Table 3.2).

Fifthly, small states have to exploit the special characteristics of their small administrations, which allow their diplomatic forces to be more informal, flexible and autonomous. The Irish delegation, during its term in the Council, proved that these traits could make a huge difference. In the aftermath of 9/11, when the US showed signs of uncertainty about taking the attack to the UNSC, the Irish delegation—informally—managed to persuade the US to do so,[105] thus strengthening the institution. The Irish case also shows that skilful negotiation tactics, competence and autonomy of officials are crucial in the Council. States need excellent negotiators in the Council to be able to exert an influence.[106] Despite the veto rights of the permanent members, the proceedings in the Council normally follow the protocol that issues should be negotiated until consensus is reached, and the elected members can use this to their advantage. Issues are usually brought to the UNSC by members of the P5 or other states precisely to garner legitimacy for proposed actions.[107] This desire for legitimacy makes it important for resolutions to be approved by consensus, which gives each UNSC member an opportunity for influence.

Sixthly, initiative taking is crucial for small states. The best opportunity for small states to do this is when holding the UNSC Presidency.[108] Small states can

---

[100]Buhaug and Voldhagen (2001).

[101]United Nations (2005a).

[102]United Nations (2005b).

[103]United Nations (2008a).

[104]Osvald, cited in Hansson (2007). See also Keating (2008), Rydberg (1998).

[105]Gillissen (2006), p. 36.

[106]Kolby (2003).

[107]Claude (1966), Voeten (2005), Hurd (2007).

[108]United Nations Security Council (2004).

use the Presidency to present a theme which is not formally on the agenda or, in the Irish case, keep things on the agenda. While the Council was caught up in 9/11, Ireland managed to maintain focus, attention and support for the peace processes in the DRC, Burundi, and Somalia.[109] In addition, during the Irish Presidency—in a context of rising violence in the Middle East following the collapse of the Oslo peace agreement—'Ireland managed to obtain unanimous agreement on a call "for immediate withdrawal of all Israeli forces" from Palestinian-ruled areas'.[110] Furthermore, Doyle argued that, despite the lack of progress, the Irish position was maintained, affirmed and decisive in shaping the overall policy of the Security Council.

Another small state, whose time on the Council is remembered for strong leadership skills and initiative it demonstrated, is New Zealand (1993–1994). New Zealand was mainly concerned with the lack of transparency in the UNSC decision-making procedure and suggested that the Council's secret deliberations should be transmitted through closed-circuit televisions so that the other UN members could be informed of, and updated on, the Council's activities. This decision did not win the hearts and minds of the permanent members.[111] However, the problem of secret consultations became apparent a few months later, at a time when New Zealand held the presidency of the Council during the Rwandan genocide. It became apparent to New Zealand's Permanent Representative that the elected UNSC members (E10) were inadequately informed about what was going on in Rwanda. The representative demanded that the E10 should be able to access information gathered by the intelligence services of the P5. When complaints were not heard, and even the Secretariat withheld information, New Zealand's Permanent Representative submitted, on his second-last day in the presidency, a draft statement to the Council that included the word 'genocide'. Rwanda—an elected member on the Council during that period—heavily objected to the statement and spread disinformation about the severity of the situation. The debate was heading towards a stalemate, so, in a last desperate attempt, New Zealand called for a draft resolution. A compromise was reached, with a watered-down statement excluding the word 'genocide'.[112] Although Rwanda represents a huge failure of the UNSC to act, without New Zealand's initiative and decisive leadership skills, no statement at all would have been released. Thus, the case represents a good example of a way in which a small state can be in pivotal position at a critical moment and make an impact.[113]

Finally, small states make use of coalition building. Having support is crucial for a small state to be able to influence the UNSC. The Nordic countries always agree

---

[109]Gillissen (2006), p. 34.

[110]Doyle cited in Gillissen (2006), p. 36.

[111]Melvern (2000).

[112]Ibid.

[113]Ross (2001).

on putting forward one of their number as a candidate for membership of the Council every other two-year term, and overcome their small size by taking joint positions and initiatives. This Nordic unity, which was highly visible during the Cold War, has in the post-Cold War era developed into greater joint European group effort at the UN.[114] There is also the CANZ group consisting of Canada, Australia and New Zealand.[115] Then, there is also the Non-Aligned Movement (NAM), consisting of 118 members.[116] According to Hurd, it has been quite common practice for certain groups of states, such as the non-aligned movement, to have extensive consultations between those members that have a Council seat and the other members in the General Assembly.[117] There is also the Forum for Small States (FOSS) which is a loose coalition of small countries which meet regularly to exchange views and coordinate positions.[118] Recognising the need for coalition building for a successful UNSC bid, Iceland established diplomatic relations with 75 countries, which is no small feat for such a small state.[119]

Furthermore, coalitions and networks with other important international actors have become crucial factors in determining whether a small state is able to influence the UNSC. The elected members have increasingly presented a thematic issue or two during the month of their presidency on topics such as disarmament, armed conflict and women's rights. If successful in bringing other actors on board, small states have a better chance of succeeding in the Council and in promoting their thematic issues. However, in order to succeed, the issue has to be well prepared, non-controversial and highly relevant.[120] There is no doubt that this networking and coalition building takes enormous time and effort and, with limited resources, small states may have to choose between presenting thematic issues within the Council or focusing on efficient participation in the daily work of the Council.[121]

3.5 Conclusion

The conventional wisdom is that the size of a state's population, territory, economy and military forces determines its influence in international organisations. Small states face political and administrative problems within the UNSC and EU decision-making processes due to their more limited resources. A lack of resources is even a problem for the richest and best-organised small states compared with the

---

[114]Laatikainen (2002).

[115]Malone (2000).

[116]British Broadcasting Corporation (2009).

[117]Hurd (2002), p. 42.

[118]Keating (2008).

[119]Gísladóttir (2008).

[120]Hansson (2007).

[121]Malone (2000).

capabilities of the large states within these organisations. Furthermore, the structural disadvantage of small states is formalised through differential voting power in the Council of the European Union whereas the structural disadvantage in the UNSC is formalised through the veto powers of the P5. There is consequently considerable doom and gloom about the prospects for small state influence in these bodies and in global politics in general.

The question arises whether small states studies have paid too much attention to quantitative factors and neglected qualitative features such as diplomatic skills and positive features related to small public administration and foreign service. Recognising that the 'traditional quantitative' factors certainly feature into a state's influence, the chapter argues that there is more to diplomacy than structural power resources. There are several additional factors that determine how states deficient in structural power can exert influence. Small states need to show the will and ambition to *try* to exert influence. Without this ambition, there will not be the long-term investment of time, effort and money needed to effectively influence proceedings. Due to limited resources, small states have to prioritise issues. Through prioritisation, small states can develop expertise and knowledge on certain topics that equals or surpasses that of large states. Influential small states have built positive images and good reputations, in part due to their successful domestic systems and dovish pasts. Small states can also make the best of their administrative capacity, by relying on the informality, flexibility, and the autonomy of officials to negotiate more effectively than diplomats from larger states. These features help small states to form networks and coalitions. Coalition building is an essential feature in determining whether small states are able to influence the EU and the UNSC.

It is therefore important to consider both the aforementioned quantitative and qualitative factors in order to produce a fuller picture of the power potential of small states in the international system. International relations scholarship cannot any longer sideline the power potential of small states within the UNSC and the EU, as is done when the focus is on measurable strengths of states in terms of number of inhabitants, size of the economy and military forces. This chapter has demonstrated that capabilities cannot only be counted in terms of these traditional variables and that special domestic features related to smallness can be used as a power tool for small states in the UNSC and the EU.

References

Archer C, Nugent N (2002) Small states and the European Union. Curr Polit Econ Eur 11:1–10
Arter D (2000) Small state influence within the EU: the case of Finland's northern dimension initiative. J Common Mark Stud 38:677–697
British Broadcasting Corporation (2009) Profile: non-aligned movement. Available at http://news.bbc.co.uk/2/hi/2798187.stm. Accessed 28 June 2016
Bourantonis D (2005) The history and politics of UN Security Council reform. Routledge, New York

Buhaug H, Voldhagen ET (2001) Norge i sikkerhetsraadet. Rapport by Utenriskdepartementet [Norwegian Foreign Department] in collaboration with NTNU. Available at http://www. regjeringen.no/en/dep/ud/dok/rapporter_planer/rapporter/2001/norge-i-sikkerhetsradet-utvidet-sikkerhe.html?id=105680. Accessed 28 June 2016

Claude I (1966) Collective legitimization as a political function of the United Nations. Int Organ 20:367–379

Danish Institute for International Studies (2005) Nordic peace diplomacy, looking back, moving forward. Background note for a Conference which was part of Norway's Centennial celebrations. Available at https://www.regjeringen.no/en/aktuelt/nordic_peace_diplomacy-looking/ id269243. Accessed 28 June 2016

Dreher A, Nunnenkamp P et al (2008) Does US aid buy UN General Assembly votes? A disaggregated analysis. Public Choice 136:139–164

Dreher A, Sturm J-E et al (2009a) Development aid and international politics: does membership on the UN Security Council influence World Bank decisions? J Dev Econ 88:1–18

Dreher A, Sturm J-E et al (2009b) Global horse trading: IMF loans for votes in the United Nations Security Council. Eur Econ Rev 53:742–757

Dreher A, Gould M et al (2013) The determinants of election to the United Nation Security Council. Public Choice 158:51–83

Duke SW (2001) Small states and European security. In: Reiter E, Gartner H (eds) Small states and alliances. Physica-Verlag, Heidelberg

European Council on Foreign Relations (2016) European foreign policy scorecard 2016. European Council on Foreign Relations, London

Flores-Macias GA, Kreps SE (2013) The foreign policy consequences of trade: China's commercial relations with Africa and Latin America, 1992–2006. J Polit 75:357–371

Gillissen C (2006) The back to the future? Ireland at the UN Security Council, 2001–2002. Nord Ir Stud 5:23–40

Gísladóttir IS (2008) Address on foreign affairs by Iceland's Minister for Foreign Affairs and External Trade. Available at www.iceland.is/iceland-abroad/jp/.../address_on_foreign_affairs_8_april_2008.doc. Accessed 28 June 2016

Government of Poland (2014) 10 PL-UE: Poland 10 Years in the European Union. Available at http://www.msz.gov.pl/resource/ef26c779-74e4-4a0c-aa73-0a9d3c8b695c:JCR.   Accessed 28 June 2016

Grön C (2014) Small states seeking influence in the European Commission: opportunities and constraints, research consortium, small state briefs. Centre for Small State Studies, Jean Monnet Centre of Excellence, University of Iceland. Available at http://ams.hi.is/wp-content/uploads/2014/04/Small-State-Briefs-5_Gron.pdf. Accessed 28 June 2016

Grön C, Wivel A (2011) Maximizing influence in the European Union after the Lisbon Treaty: from small state policy to smart state strategy. J Eur Integr 33:523–539

Gylfason ÞF (2004) The distinctive characteristics and behaviour of small states which have been elected members to the United Nations Security Council. Thesis (BA), University of Iceland

Habeeb W (1988) Power and tactics in international negotiations: how weak nations bargain with strong nations. Johns Hopkins University Press, Baltimore

Hansson P (2007) Against all odds: small state candidature and membership of the United Nations Security Council. Thesis (MA), University of Iceland

Hulton SC (2004) Council working methods and procedures. In: Malone D, Boulder DM (eds) The UN Security Council: from the Cold War to the 21st century. Lynne Rienner Publisher, London

Human Development Report (HDR) (2007/2008) Human development index 2005: monitoring human development, enlarging people's choices. Fighting climate change: human solidarity in a divided world

Human Development Report (HDR) (2009) Human development index 2007 and its components. Overcoming barriers

Hurd I (2002) Legitimacy, power, and the symbolic life of the UN Security Council. Glob Gov 8:35–51

Hurd I (2007) After anarchy: legitimacy and power in the United Nations Security Council. Princeton University Press, Princeton

Ikenberry GJ (2001) After victory: institutions, strategic restraint and the rebuilding of order after major wars. Princeton University Press, Princeton

Ingebritsen C (2006) Norm entrepreneurs: Scandinavia's role in international relations. In: Ingebritsen C, Neumann I et al (eds) Small states in international relations. University of Washington Press/University of Iceland Press, Seattle/Reykjavik

Jakobsen PV (2009) Small states, big influence: the overlooked nordic influence on the civilian ESDP. J Common Mark Stud 47:81–10

Katzenstein P (2003) Small states and small states revisited. New Polit Econ 8:9–30

Keating C (2008) The United Nations Security Council: options for small states. Speech by the Executive Director, Security Council Report. Available at www.securitycouncilreport.org/atf/cf/%7B65BFCF9B-6D27-4E9C8CD3CF6E4FF96FF9%7D/Media%20Small%20States%20Reykjavik.pdf. Accessed 28 June 2016

Keohane RO (1969) Lilliputians' dilemmas: small states in international politics. Int Organ 23:291–310

Keohane RO, Martin L (1995) The promise of an institutional theory. Int Secur 20:39–51

Kolby OP (2003) Norge i FNs Sikkerhetsrad 2001–2002. Internasjonal Politikk 61:277–287

Krisch N (2008) The security council and the great powers. In: Lowe W, Roberts A et al (eds) The United Nations Security Council and war: the evolution of thought and practice since 1945. University Press Oxford, Oxford

Kronsell A (2002) Can small states influence EU norms? Insights from Sweden's participation in the field of environmental politics. Scand Stud 74:287–304

Kuziemko I, Werker E (2006) How much is a seat on the Security Council worth? Foreign aid and bribery at the United Nations. J Polit Econ 114:905–930

Laatikainen KV (2002) Norden's eclipse: the impact of the European Union's common foreign and security policy on the nordic group in the United Nations. Coop Conf 38:409–441

Lim DYM, Vreeland JR (2013) Regional organisations and international politics: Japanese influence over the Asian Development Bank and the UN Security Council. World Polit 65:34–72

Lægreid P (2000) Implications of europeanization on central administration in the nordic countries. Paper presented at IASIA Annual Conference, Working Group III: Public Service Reform, Beijing, 10–13 July 2000

Maes I, Verdun A (2005) Small states and the creation of the EMU: Belgium and the Netherlands, pace-setters and gate-keepers. J Common Mark Stud 43:327–348

Magnúsdóttir GL (2009) Small states' power resources in EU negotiations: the case of Sweden, Denmark and Finland in the environmental policy of the EU. Thesis (PhD), University of Iceland

Malone D (2000) Eyes on the prize. Glob Gover A Rev Multilateralism Int Org 6:1–21

Mearsheimer JJ (1995) The false promise of international institutions. Int Secur 19:5–49

Mearsheimer JJ (2001) The tragedy of great power politics. Norton, New York

Melvern L (2000) Behind closed doors. World Today 56:9–11

Morgenthau HJ (1972) Politics among nations, 5th edn. Alfred A. Knopf, New York

Novosad P, Werker E (2014) Who runs the international system? Power and the staffing of the United Nations Secretariat. Center for Global Development Working Paper 376. Available at www.cgdev.org/publication/who-runs-international-system-power-and-staff-united-nations-secretariat-working. Accessed 28 June 2016

Nye JS (1990) Soft power. Foreign Policy 80:153–171

Odén B (2010) Review essay. Forum Dev Stud 37:269–279

Panke D (2010) Small states in the European Union: coping with structural disadvantages. Ashgate, Farnham

Petersen FA (1998) The international situation and Danish foreign policy 1997. In: Heurling B, Mouritzen H (eds) Danish foreign policy yearbook, 1998. Danish Institute of International Affairs, Copenhagen

Ross K (2001) Power in numbers. World Today 57(5):23–24

Rydberg R (1998) Sverige i Sakerhetsradet-arbete och organisation i Stockholm. In: Promemoria 1998-12-28. Regjeringskansliet Utenriksdepartementet, Stockholm

Schweller RL (2003) The progressiveness of neoclassical realism. In: Elman C, Elman MF (eds) Progress in international relations theory: appraising the field. MIT Press, Cambridge

Tallberg J (2008) Bargaining power in the European Council. J Common Mark Stud 46:685–708

Thorhallsson B (2000) The role of small states in the European Union. Ashgate, Burlington

Thorhallsson B (2004) Can small states influence policy in an EU of 25 members? In: Busek E, Hummer W (eds) Liechtenstein Politische Schriften, vol 39. Verlag der Liechtensteinischen Akademischen Gesellschaft

Thorhallsson B (2006a) The size of states in the European Union: theoretical and conceptual perspectives. J Eur Integr 28:7–31

Thorhallsson B (2006b) Iceland's involvement in global affairs since the mid-1990s: what features determine the size of a state. Icel Rev Polit Adm 2:197–223

Thorhallsson B (2012) Small states in the UN Security Council: means of influence? Hague J Dipl 7:135–160

Thorhallsson B (2013) The Icelandic economic collapse: how to overcome constraints associated with smallness? Eur Polit Sci 12:320–332

Thorhallsson B, Wivel A (2006) Small states in the European Union: what do we know and what would we like to know? Camb Rev Int Aff 19:651–668

Toplak C (2014) Small states in Europe: Slovenia and its position within the European Union. Lecture at the Centre for Small States Studies, University of Iceland, 27 October 2014

Ulriksen S (2006) Deployments for development? Nordic peacekeeping efforts in Africa. Norwegian Institute of International Affairs Publication, Oslo

United Nations (1983) Provisional rules of procedure of the Security Council. UN Doc S/96/Rev. 7

United Nations (2005a) Press release: effective protection of civilians in armed conflict requires stronger partnerships. Available at www.un.org/News/Press/docs/2005/sc8575.doc.htm. Accessed 28 June 2016

United Nations (2005b) Press release: Security Council stresses urgency of full, effective implementation of 'landmark? Resolution 1325 (2000) on women, peace and security. Available at www.un.org/News/Press/docs/2005/sc8538.doc.htm. Accessed 28 June 2016

United Nations (2008a) Press release: Security Council hears 60 speakers, asks Secretary-General to advise organization within one year of best ways to support national peacebuilding efforts. Available at www.un.org/News/Press/docs/2008/sc9333.doc.htm. Accessed 28 June 2016

United Nations (2008b) Biographical note: Secretary-General appoints Kai Eide of Norway as special representative for Afghanistan. Available at www.un.org/News/Press/docs/2008/sga1123.doc.htm. Accessed 28 June 2016

United Nations (2010a) Contributors to United Nations Peacekeeping Operations – Aug 2010. Peacekeeping Department, New York Available at http://www.un.org/en/peacekeeping/resources/statistics/contributors_archive.shtml. Accessed 28 June 2016

United Nations (2010b) Biographical note: Secretary-General appoints Staffan de Mistura of Sweden as special representative for Afghanistan. Available at www.un.org/News/Press/docs/2010/sga1217.doc.htm. Accessed 28 June 2016

United States Census Bureau (2010) Countries and Areas Ranked by Population. Available at http://www.census.gov/population/international/data/countryrank/rank.php. Accessed 28 June 2016

United Nations Security Council (2003) Repertoire of the practice of the Security Council 2000–2003. Participation in the proceedings of the Security Council, Chapter 3. Available at www.un.org/en/sc/repertoire/. Accessed 28 June 2016

United Nations Security Council (2004) United Nation Security Council. UN Doc S/2004/135

United Nations Security Council (2007) Repertoire of the practice of the Security Council 2004–2007. Participation in the proceedings of the Security Council, Chapter 3. Available at www.un.org/en/sc/repertoire/. Accessed 28 June 2016

Utanríkisráðuneytið (2009) Skýrsla um framboð Íslands og kosningabaráttu til sætis í öryggisráði Sameinuðu þjóðanna, 2009–2010. Available at https://www.utanrikisraduneyti.is/media/PDF/Lokaskyrsla_um_oryggisradsframbodid_2008.PDF. Accessed 28 June 2016

Van den Berg DJ (1994) The Netherlands and Luxembourg: smaller countries in an ever-larger Europe. EISPASCOPE 3:1–4

Vital D (1967) The inequality of states: a study of small power in international relations. Clarendon Press, Oxford

Voeten E (2005) The political origins of the UN Security Council's ability to legitimize the use of force. Int Organ 59:527–557

Vreeland JR, Dreher A (2014) The political economy of the United Nations Security Council money and influence. Cambridge University Press, New York

Waltz KN (1959) Man, the state, and war. Columbia University Press, New York

Waltz KN (1979) Theory of international politics. McGraw-Hill, Reading

Whitfield T (2007) Friends indeed? The United Nations, groups of friends, and the resolution of conflict. US Institute of Peace Press, Washington

Wight M (1991) The three traditions. In: Porter B, Wight G (eds) International theory: the three traditions. Leicester University Press, Leicester

Wivel A (2005) The security challenge of small EU member states: interests, identity and the development of the EU as a security actor. J Common Mark Stud 3:393–412

Chapter 4
The Impact of EU Law in Luxembourg: Does Size Matter?

Michèle Finck

4.1 Introduction

This chapter engages with Luxembourg's nature as a small state, and the impact its size may have on its relation with the European Union (and in particular its legal order), of which it is a founding Member State. When it comes to size, Luxembourg's relationship to the European Union is ambiguous. Territorially and demographically speaking, Luxembourg is, with its 2.586 km^2 and just over half a million inhabitants, a lightweight, no doubt.[1] Historically and politically speaking, Luxembourg however punches above its geographical weight. It is not only a founding Member State of the EU but has also continuously acted and been perceived as a loyal partner of the European integration project, a stable economy embedded in the internal market and a significant diplomatic player. The European Commission and the institutions that preceded it have had eighteen presidents over time, three of which were Luxembourg nationals: Gaston Thorn was the President of the European Commission from 1981–1985, Jacques Santer from 1995–1999 and Jean-Claude Juncker is the Commission's current head. Luxembourg, and especially its capital, Luxembourg City, host divisions of the European Parliament and the European Commission as well as the Court of Justice of the European Union ('CJEU' or 'the Court'). In April, June and October of each year, the meetings of the Council are held in Luxembourg.[2]

---

[1]European Union (2016).

[2]European Council in Edinburgh (1992): While the seat of the European Council is in Brussels, it was decided at the European Council in Edinburgh in 1992 that during these 3 months, the European Council would convene in the Grand Duchy.

M. Finck (✉)
Department of Law, London School of Economics, London, UK
e-mail: m.finck@lse.ac.uk

© Springer International Publishing AG 2017
P. Butler, C. Morris (eds.), *Small States in a Legal World*,
The World of Small States 1, DOI 10.1007/978-3-319-39366-7_4

    Luxembourg has also been denoted as one of the, if not the, most Euro-friendly
Member States. In 1986, the prestigious 'Karlspreis' (or 'Prix Charlemagne') that
honours those that make considerable advancements to European integration, was
awarded to the people of Luxembourg.[3] It is, to this date, the only Member State to
have been awarded this prize. The citation for the award interestingly alludes to
Luxembourg's size: '[w]ith this award the people of the smallest country in the
European Community were honoured for the fact that they were convinced
Europeans from the outset and their politicians have made important contributions
to the European Union'.[4] This sums up that despite its small size, the Grand Duchy
has long been considered a key actor in the furthering of European integration.
    The preceding brief introductory overview underscores that on the one hand
Luxembourg is the second smallest Member State of the EU, both by population
and territorial size (only Malta has a noticeably smaller territory and a slightly
smaller population).[5] On the other hand, however, Luxembourg has been an active
proponent of European integration, especially during the early stages of the
EU. These facts do not necessarily correlate, but they might. Politically speaking,
the relative importance of Luxembourg in the EU can certainly be explained by a
number of elements, such as its nature as a founding Member State, but also its size.
It can indeed be assumed that one of the factors that led other Member States to
agree to having a Luxembourger head the European Commission is precisely
because of the Grand Duchy's small size. In this scenario, there is no risk of
creating the impression that Luxembourg is becoming 'too' powerful in the EU
through this presidency. Things would be very different if, say, a German national
were the President of the Commission. This brief historical overview of
Luxembourg's political role in the crafting of the European Union indicates that
size might matter when we try to grasp the role the Grand Duchy has had in this
regard. This is not however the main focus on this chapter, which is predominantly
interested in law rather than politics. The chapter's aim is to verify whether
smallness constitutes a useful lens through which to look at the Grand Duchy's
relation with the European Union.
    While it is well known that the interaction between domestic (constitutional)
legal orders and supranational law has created many intriguing legal dynamics that
have been extensively documented in the existing scholarly literature, my interest
consists in verifying whether territorial size can be considered to be a factor
impacting on the interaction between the domestic and EU legal systems. The
aim of my enquiry in the context of this book on small states is to determine
whether smallness can have an impact on a state's relation with a supranational
legal order it is a member of. My analysis is structured as follows. In providing an
overview of EU law's status in Luxembourg, I will first look at the reception of the
doctrine of supremacy in the Grand Duchy. Subsequently I will examine

---

[3]Der Internationale Karlspreis zu Aachen (2016a).

[4]Der Internationale Karlspreis zu Aachen (2016b).

[5]European Union (2016).

Luxembourg's reform of political rights as a consequence of European integration and then look at a recent case that arose in the context of the free movement of workers to determine whether this principle impacts differently on small Member States. Finally, I will introduce recent debates on fundamental rights in the Luxembourg context and suggest that these debates might have played out differently in larger Member States with a more pronounced critical mass of commentators. In the concluding section I will argue that it is fair to assume that Luxembourg's small territorial size in fact impacts on its relation with the supranational legal order.

4.2   The Status of EU Law in Luxembourg

This section, which constitutes the chapter's core, examines a consequence of EU integration that is of particular relevance from a legal perspective, namely the incorporation of EU law into Luxembourg's legal order. A few aspects of the relation between Luxembourg law and the law of the EU will be explored. First, the reception of the doctrine of supremacy in Luxembourg; second, the reform of the domestic voting rights regime as a consequence of European integration; third whether certain provisions of EU law affect Member States differently depending on their size; and finally the impact of EU law on Luxembourg's reform of its constitutional fundamental rights regime.

4.2.1   The Reception of the Doctrine of Supremacy in Luxembourg

Two legal doctrines form the pillars of EU law and its constitutional development: supremacy and direct effect. According to the doctrine of supremacy (also referred to as 'primacy'), first formulated in *Costa v ENEL* and most powerfully affirmed in *Internationale Handelsgesellschaft*, a norm of EU law takes precedence over any conflicting national law, whatever their respective status in the hierarchy of norms, and notwithstanding the moment in time at which they were issued.[6] The fact that EU law unequivocally takes precedence over conflicting domestic law in this manner has ensured the success and efficiency of EU law over time.[7] The doctrine of direct effect, first formulated in *Van Gend en Loos*, holds that EU law may, under certain conditions, confer rights on individuals in the EU's various Member States. This allows individuals to immediately invoke EU law before a domestic Court, independently of whether the EU norm has been transposed into the domestic legal

---

[6] 6/64 [1964] ECR 585; 11/70 [1970] ECR 1125.
[7] For scholarly commentary see de Witte (2011); Kumm (2005); von Bogdandy and Schill (2011).

order.[8] Just as supremacy, direct effect ensures the application and effectiveness of EU law within the various Member States.

This section focuses on supremacy. One of the parameters conventionally employed by EU lawyers to assess the early success of EU law within the different Member States is to examine the degree and ease with which supremacy was embraced by national judges.[9] This is particularly interesting with regard to the EU's founding Member States that were rightly surprised by the creation of the doctrine of supremacy by the CJEU.[10] The founding Treaties indeed contained no reference to these doctrines and it was through an exercise of teleological interpretation that they were formulated by the Union's highest Court in 1963-64.[11] As a result, opposition to the doctrines surfaced in a number of contexts, most famously in the highest domestic jurisdictions of Germany and Italy.[12] To a degree, the resulting constitutional debates are still on-going in some Member States.[13] In Luxembourg, however, supremacy has never been subject to any genuine controversy. This section explores the reasons underlying this fact, and enquires whether they relate to its smallness.

4.2.1.1   A Brief History of Constitutional Independence

One of the reasons why supremacy was opposed in other Member States is the doctrine's inherent ability to undermine national constitutional autonomy. If the content of domestic constitutional law must comply with EU law, national constituents can, notwithstanding formal sovereignty, no longer do as they please, as their rules might be overridden by those of EU law, and indeed even EU secondary legislation. This evolution naturally caused concern in those states with a well-established history of national and constitutional independence as supremacy signalled a reversal of a well-established status quo. It also proved problematic in Germany, where the post WWII Constitution and its fundamental rights regime have a particular domestic significance; supremacy was considered a threat to their strength and perpetuity. In Luxembourg, however, the situation was different. Prior to joining the EU Luxembourg had only known a comparatively brief period of constitutional independence, and its existing constitutional order was already

---

[8]26/62 [1963] ECR.

[9]Craig and de Búrca (2011), pp. 268–296.

[10]In contrast, those Member States that joined at a later stage knew that supremacy and direct effect were part of EU law.

[11]Treaty of Lisbon 2007, Declaration Number 17, 12008E/AFI/DCL/17. Now, Declaration No. 17: concerning primacy enshrines the principle of supremacy, as formulated by the Court.

[12]Craig and de Búrca (2011), pp. 272–285.

[13]*R (on the application of HS2 Action Alliance Limited) v Secretary of State for Transport and another* [2014] UKSC 3.

characterised by a particular openness towards international law. A brief enquiry into Luxembourg's history serves to make this point.

Luxembourg was founded in 963.[14] For most of its history, however, the Grand Duchy was not an independent State with its own Constitution. It rather belonged successively to Burgundy, Spain, France, Austria, the Netherlands, and Germany, sometimes being handed back and forth between them.[15] Luxembourg's existence as an independent State only dates back to the second half of the nineteenth century. The 1815 Congress of Vienna finally gave formal authority to Luxembourg as an independent State.[16] Its own Constitution, however, was only granted in 1841[17] and was quickly replaced by a revised constitutional document in 1868.[18] The 1868 Constitution is still in force today, at least for the time being, as a constitutional revision, which is introduced in further depth below, is currently underway.[19]

Article One of the Constitution of 1868, still in force at the present moment in time, for the first time defined the Grand Duchy as an 'independent' State—less than a 100 years before the start of European integration. Significantly, there were also periods between 1868 and the beginnings of European integration during which Luxembourg was robbed of its independence. The Grand Duchy was occupied by German troops during WWI and was formally annexed to Germany in WWII. Before and after the two World Wars, Luxembourg was a member of transnational alliances. In 1842 it joined the German Zollverein, and in 1921 it entered into an economic and monetary union with Belgium, the Union Economique Belgo-Luxembourgeoise (UEBL).[20] Following the Second World War, Luxembourg joined the BeNeLux, an economic union with Belgium and the Netherlands, and helped found the European Union.[21]

This nutshell account of Luxembourg's history underlines that the country has not known any extended periods of complete independence. This is noteworthy for the following reason. Prior to 1958, when the Treaty of Rome establishing the European Economic Community (the predecessor of what is now the EU) was signed, Luxembourg had known only a relatively brief period of constitutional

---

[14]Luxembourg's recorded history can, however, be traced back to Roman times.

[15]Gerkrath and Thill (2012), p. 1087.

[16]Its independence was however only formally ratified in 1867, the date until the Dutch King retained sovereignty over Luxembourg. For an overview of Luxembourg's history, see Trausch (1992); Pauly (2013).

[17]For an overview of Luxembourg's constitutional history, see Gerkrath and Thill (2012).

[18]Weber (1957); Ravarani (2007), pp. 59–83.

[19]For an overview of this process in English, see Gerkrath (2013a).

[20]Luxembourg also joined the League of Nations in 1920.

[21]The creation of the BeNeLux was decided in 1943 when the governments of Luxembourg, Belgium and the Netherlands were in exile in London. It was then formally founded by the BeNeLux Treaty of 1958, which was signed for a duration of 50 years (and renewed in 2008). The Treaty of Rome of 1958 founded the European Economic Community, the predecessor of what currently is the European Union.

independence, after a long history of forming part of larger units. This is of course a very different historical background from EU Member States, such as, for instance, the United Kingdom with its notion of parliamentary sovereignty, which has been reaffirmed over the course of centuries.[22] Being subject to rules imposed on it via 'the outside', in this case the EU, was not new to Luxembourg. If anything it could be argued that Luxembourg's independence—at least in a certain sense—increased with European integration. Contrary to previous periods of occupation and submission, Luxembourg now had a say in the rules that it would be subject to as, as a EU Member State, it was one of the masters of the EU Treaties and could determine the shape of primary and secondary legislation.[23] It can further be speculated that the European integration process, aimed at creating peace in Europe and thus ending war and occupation between Member States, created a sense of security and independence that the Grand Duchy did not benefit from when it was but a small state stuck between war-waging larger states and in constant risk of being occupied. These various factors may explain the smooth reception of the doctrine of supremacy in Luxembourg domestic law, which is outlined just below.

4.2.1.2   Luxembourg and the Supremacy of EU Law

The reception of the doctrine of supremacy in the Luxembourg domestic legal order was genuinely uneventful.[24] It is important to note in this respect that the formalisation of the supranational project was preceded by a domestic constitutional amendment that formalised Luxembourg's welcoming approach towards European integration. In 1956, after the European Coal and Steel Community had been created,[25] but prior to the signing of the Treaty of Rome, Luxembourg amended its 1868 Constitution to reflect its membership of said supranational organisations.[26] This constitutional revision was considered necessary as the organs of these supranational organisations, and of the EU nowadays, exercised powers 'which the Luxembourg Constitution conferred on the legislature, executive and

---

[22]According to AV Dicey, 'The principle of parliamentary sovereignty means neither more nor less than this: namely that parliament thus defined has, under the English constitution, the right to make or un make any law whatever; and further, that no person or body is recognized by the law of England having a right to override or set aside the legislation of parliament...'. See Dicey (1915), pp. 3–4.

[23]Article 48 TEU governs the Treaty revision procedure in the EU and Article 294 TFEU outlines the EU's Ordinary Legislative Procedure.

[24]This mirrors the peaceful reception of international Treaties in EU law.

[25]The European Coal and Steel Community was the predecessor of the European Economic Community. It was created by the 1951 Treaty of Paris.

[26]Such revision was considered all the more necessary as the ECSC was a success and that the formation of similar communities was being considered, see Gerkrath and Thill (2012), p. 1092.

judiciary'.[27] Article 49b of the Luxembourg Constitution allows these powers to be transferred to the supranational level. It is phrased as follows:

> The exercise of powers reserved by the Constitution to the legislature, executive and judiciary may temporarily be vested by treaty in institutions governed by international law.[28]

Nowadays, the fact that this transfer is solely of a 'temporary' nature is subject to controversy and is being rethought in light of the current constitutional revision.[29] After the insertion of Article 49b into the Luxembourg Constitution, Courts also took a welcoming approach towards the supremacy of EU law.[30] Despite the Constitution's silence regarding the status of EU law vis-à-vis conflicting domestic provisions, Courts and scholars had long admitted the primacy of Treaty law over internal law, including the constitution.[31] Luxembourg Courts have endorsed the view that domestic legislation, even that posterior in time to the Treaty in question, should not be applied where it conflicts with a directly applicable Treaty provision. According to Mak, 'Luxembourg courts accepted and applied the CJEU's judgments concerning the nature of the EC legal order and the supremacy of EC law in the legal orders of the member states'.[32] While the Constitution is to this date silent on the matter, the acceptance of supremacy has been reiterated over time by various organs.[33] In 1984 the Grand Duchy's Council of State stressed the EU's special nature as a 'new legal order', echoing the CJEU's famous lines in *Van Gend en Loos*, in stating that:

> Where an internal norm conflicts with a norm of international law having direct effects in the domestic legal order, Treaty law must prevail. This rule applies in particular where the conflict arises between a norm of domestic law and a norm of EU law as the European Treaties have created a new legal order for the benefit of which Member States have limited the exercise of their sovereign powers in the areas determined by the Treaties.[34]

---

[27] Such revision was considered all the more necessary as the ECSC was a success and that the formation of similar communities was being considered, see Gerkrath and Thill (2012), p. 1092.

[28] Constitution of Luxembourg 1868, art 37. This is my own translation. According to the simultaneous modification of article 37, such treaties have to be approved by a law adopted under the same majority requirements as those, which apply for constitutional revision.

[29] Article 5 of the Draft Constitution provides that the Grand Duchy 'participates in the European Union' and that 'the exercise of the State's powers may be transferred to the European Union [...]' by an act passed by qualified majority'. The fact that the transfer of powers to the EU, as currently envisaged by the Luxembourg Constitution is only temporary and has been subject to criticism by Gerkrath (2013a), p. 459.

[30] Moyse (2005).

[31] An answer to parliamentary question n° 1538 from June 21, 2011 issued by MP Alex Bodry. On this, see Gerkrath (2013b). The statute of the Luxembourg Constitutional Court also excludes international Treaties from constitutional review.

[32] Mak (2010), p. 322.

[33] Mak (2010), pp. 311, 322; Spielmann (2001), p. 532; Cour supérieure de justice (cassation criminelle), 14 July 1954, Pas. lux., XVI, 150, *Chambre des Métiers v Pagani*; Conseil d'Etat, 28 July 1951, Pas. lux. XV, 263, *Dieudonné v Administration des Contributions*.

[34] Conseil d'Etat, 21 November 1984, Pas. lux., XXVI, p. 174. Translation by author.

In its 1992 advisory opinion on a planned constitutional amendment to allow non-Luxembourgers to participate in local and European elections (a requirement arising under EU law, which is examined in further detail below) the Council of State moreover stated that:

> according to the hierarchy of rules of law, international law takes precedence over national law and that in the event of conflict the courts disapply internal law in favour of the Treaty. As it is desirable that there be no contradiction between our national law and international law, the Council of State demands that the necessary amendment to the Constitution be made in good time so as to avoid such a situation of incompatibility.[35]

The primacy of EU law is hence uncontested in Luxembourg, even if the Constitution is mum on the issue and the Constitutional Court, created in only 1996, never explicitly pronounced itself on the matter.[36] This might be traced back to the absence of any controversy on the matter. According to Gerkrath, 'Luxembourg's courts have had no difficulty in recognising the pre-eminence of international law and the primacy of Community Law, including in respect of a constitutional provision'.[37] This judicial position is also echoed widely in existing legal scholarship. While scholarship engaging with Luxembourg constitutional law is necessarily limited, a point I will come back to further below, a survey undertaken by Mak has concluded that not a single scholarly criticism of the supremacy of EU law can be pinned down.[38]

While Article 49b can be considered as an element easing such reception, Gerkrath has also noted that this may be due to a particular characteristic of the Luxembourg legal order, precisely that, 'unlike other European countries, it is founded on international law'.[39] This echoes the absence of a long period of constitutional independence outlined above. The Grand Duchy was indeed established 'as an independent state by the Final Act of the Congress of Vienna of 9 June 1815', an independence that was confirmed by the Treaties of London of 1839 and 1867.[40] It follows that 'far from constituting a threat to national sovereignty' international law is understood as a guarantee of the 'existence and survival of the state'.[41] This indicates that rather than seeing its autonomy threatened by transnational and supranational integration, the European Union has in fact fortified the independence Luxembourg gained shortly before its beginnings. The supremacy of EU law over conflicting domestic law is in any case well-established in Luxembourg and has not been subject to any noteworthy controversy, very much to

---

[35]European Justice (2016).

[36]The Constitution of Luxembourg 1868, art 95: This article was introduced by the constitutional revision of 12 July 1996, and the Constitutional Court Act 1997.

[37]Gerkrath (2014), p. 123; Pescatore (1962), p. 96; Kinsch (2010), pp. 383–399.

[38]Mak (2010), p. 320.

[39]Gerkrath (2014), p. 123.

[40]Ibid.

[41]Ibid.

the contrary of many other Member States. We must now enquire whether this state of affairs has any link to the country's size.

4.2.1.3   The Impact of Smallness

This section has thus far established that constitutional independence came to Luxembourg relatively shortly before European integration, and that the supremacy of EU law was well-received in Luxembourg. This short sub-section enquires whether Luxembourg's smallness has had any significance regarding these matters. While a number of factors could certainly be invoked to explain why Luxembourg has not opposed the supremacy of EU law, I will put forward the hypothesis that its smallness can indeed be seen to play a role in this regard. It can in fact be speculated that the very reason why Luxembourg had no firm history of entrenched constitutional independence is its small size.[42] Due to its small size, and a correspondingly small army, it was straightforward for other states to take control of Luxembourg. Indeed, as the statement by Gerkrath above underlines, the Grand Duchy needed the support and protection of international law to establish and protect its independence. As a small country, it could not do this on its own. This would mean that, because Luxembourg is a small state, it has no long history of independence, a fact that in turn facilitated the reception of EU law.[43] We can thus assume that Luxembourg's geographical scale has had an impact on its relation to EU law, and thus European integration when it comes to supremacy. The next section examines whether the same can be said regarding the reframing of domestic law in light of EU citizenship.

4.3   The Reform of Political Rights as a Result of European Integration

The legal orders of all Member States have been through deep substantive modifications as a result of European integration. This encompasses most if not all substantive areas of law. This section examines how European integration has affected one particular aspect of Luxembourg law, namely its voting legislation. Traditionally, the active and passive right to vote were restricted to those having a Luxembourg passport. This echoed the situation prevailing in the large majority of

---

[42]Other factors matter too in this regard, certainly for instance Luxembourg's geographical location.

[43]Of course there may be many other reasons. It would for instance be interesting to study whether the fact that the CJEU is located in Luxembourg, and that its judgements are rendered in French, the lingua franca of Luxembourg law, have any role to play.

EU Member States at the time.[44] The Maastricht Treaty however gave birth to the notion of European citizenship, which entails a limited number of concrete rights for EU citizens, that is to say the nationals of any Member State of the EU.[45] These rights include the ability for European citizens to participate in local elections and elections for the European Parliament in their state of residence, even if they are not nationals of that state.[46] In light of these provisions, Luxembourg has had to modify its voting legislation and also the relevant provisions of its Constitution in order to align them with requirements arising under supranational law.[47]

While, as a matter of principle, European citizenship has the same consequences for all EU Member States, a peculiarity arose with regard to Luxembourg, which led to the application of a different legal regime to the Grand Duchy. To understand this peculiarity, it is necessary to first be clear on Luxembourg's demographic structure. Luxembourg has almost 563,000 inhabitants, of whom 45.9% do not have Luxembourg nationality.[48] Over the past century Luxembourg's population doubled as the country went from being a mostly rural and then subsequently industrial society to one of the world's leading banking hubs.[49] This rapid economic growth significantly broadened the country's labour market, and foreign workers were needed both because there were and are simply not enough nationals to fill all vacancies and because they did not always have the required skills. Nowadays foreign residents and cross-border workers make up 71% of the country's work-force.[50] Despite 150 nationalities being represented in the Grand Duchy, 86% of all foreign residents are European citizens, that is to say nationals from Member States of the EU other than Luxembourg.[51]

Recognising that in light of its particular demographic composition the practical impact of the Maastricht citizenship reform would be of a different nature in Luxembourg than in all other Member States, a special regime was created under EU law, which de facto only applies to the Grand Duchy. As per this regime Member States with a high share of foreign residents can impose conditions on EU national's ability to participate in EU and local elections. Article 14 of Directive

---

[44]Lardy (1997), p. 75. An exception in this regard is for instance the UK, where Irish nationals and 'qualifying' Commonwealth citizens can vote. On this, see Representation of the People Act 1983, s 1(1)(c).

[45]For an overview of the concept of EU citizenship, see by way of example works by Francis Jacobs.

[46]TFEU, art 22; Council Directive 93/109/EC OJ 1993 L329/34; Council Directive 94/80/EC OJ 1994 L368/38. This right has also been enshrined in the Charter of Fundamental Rights of the European Union, namely arts 39(1) & 40.

[47]Constitution of Luxembourg 1868, art 107: which amended by the constitutional revision of 23 December 1994. See also Gerkrath (2013b), p. 544.

[48]Institut national de la statistique et des études économiques (2015).

[49]Ibid. In 1910, Luxembourg had about 260,000 inhabitants, most of Luxembourg nationality.

[50]Government of the Grand Duchy of Luxembourg (2015). There are also 10,000 people employed by the EU institutions in Luxembourg.

[51]Index Mundi (2015); Institut national de la statistique et des études économiques (2015).

93/109/EC and Article 12(1) of Directive 94/80/EC provide that if a Member State's population accounted for more than 20 per cent of non-national EU citizens of voting age in 1996, then this state may restrict their right to vote in European and/or local elections. Luxembourg was the sole Member State fulfilling these criteria. During an initial period, the Grand Duchy made full use of these derogatory mechanisms to make participation in elections conditional subject to a number of conditions for EU citizens and hence restrict their involvement in the democratic process. By now, the Grand Duchy has however chosen to no longer rely on these derogations with regard to European elections. It however continues to apply them to local elections, which explains why foreigners must reside in Luxembourg during at least 5 years before they can cast their ballot.[52] Luxembourg has thus modified its domestic electoral law in light of EU law, but has been able to do so in a manner different from that of all other Member States. The relevant question for the purposes of this chapter is whether this evolution can be connected to Luxembourg's territorial size.

Lansbergen and Shaw illustrated that via the existing exemptions, EU law explicitly recognises that 'the very fact that EU citizens are present in large numbers and could present, therefore, a significant challenge to the system has been sufficient reason to deny, effectively, the implementation of Article 19 EC [now Article 22 TFEU]'.[53] The exemption is unequivocally based on the percentage of foreign residents in a Member State—and in Luxembourg this percentage probably relates to its small size. It is in fact difficult to imagine that Luxembourg would have an equally high percentage of foreign residents if it were not for its small overall population. The predominant reason explaining why Luxembourg has attracted and continues to attract foreign workers in large numbers is its still thriving economy, which relies on foreign workers to function. People come to Luxembourg to work and because the Luxembourg economy heavily depends on them they make up a large share of the overall population.[54] It is however difficult to imagine that, had Luxembourg a higher number of inhabitants to start with, say ten million, foreign workers and their relatives could make up an equally significant share of the population. Indeed, while other small countries, such as Qatar or Singapore also have equally high shares of foreign workers, there appears to be no precedent of larger countries being subject to such dynamics.

Yet it was the single fact that Luxembourg has such a high share of foreigners among its population that justified the derogation created under EU law, which entails that European citizenship has different legal consequences in Luxembourg than it has in any other Member State. Smallness thus appears to matter and

---

[52]Loi électorale du 18 février 2003, art 2(4).

[53]Lansbergen and Shaw (2010), p. 62.

[54]Luxembourg's demographic structure and labour economics are certainly more complex than these general statements indicate. They are nonetheless indicative of a broad pattern.

impact on the effects EU law has in Luxembourg's domestic legal order. As Gerkrath noted:

> Union membership modifies the meaning of many important concepts of constitutional law. This phenomenon of Europeanization can be observed in every member state, and has been particularly highlighted on the occasion of recent accessions. It appears even more obviously in a small founder state such as the Grand Duchy whose commitment to the process of European integration needs no further demonstration.[55]

While Gerkrath did not elaborate further on the relation between EU citizenship and Luxembourg's nature as a small state, his reference to the latter characteristic nonetheless indicates an intuition that smallness has an impact on the reception of EU law in Luxembourg, particularly in the context of voting rights. The impact of smallness does not however stop here. A similar impact can also be perceived in the context of recent debates on electoral reform. Precisely because of this high share of foreign residents, a debate emerged in the past years regarding democratic legitimacy of a vote made up solely by small majority of the overall population. The main argument was that, as almost 50% of those living in the country cannot participate in national legislative elections, the Luxembourg democracy should aim to be more inclusive. After a change in political power-constellations and as a new government entered office[56] plans emerged to hold a referendum on the question whether foreign residents should be able to participate in national legislative elections.[57] Had this been successful, Luxembourg would have become the first EU Member State to proceed to such a reform, again triggered by a demographic composition that at least in part results from smallness.[58]

After much debate the question that was put to the vote was whether foreign residents should participate in parliamentary elections subject to two conditions: first, that the individual has resided in Luxembourg for at least 10 years, and second, that she previously participated in municipal or European elections.[59] While initial polls suggested that up to 59% percent of Luxembourgers favoured such reform the final result looked very different.[60] As a matter of fact, a sobering 78.02% percent voted against the proposal. There is no space here to engage in additional depth with this

---

[55]Gerkrath (2014), p. 111.

[56]This was the first government in a few decades that the Conservative Party formed no part of.

[57]On the Luxembourg referendum, see Heuschling (2013); Gerkrath (2013a).

[58]I develop this argument further in Finck Michèle Finck 'Towards an Ever Closer Union Between Residents and Citizens? On The Possible Extension of Voting Rights to Foreign Residents in Luxembourg' 11 European Constitutional Law Review (2015) 78-98.

[59]The French text of the relevant referendum question read as follows: 'Approuvez-vous l'idée que les résidents non luxembourgeois aient le droit de s'inscrire de manière facultative sur les listes électorales en vue de participer comme électeurs aux élections à la Chambre des Députés, à la double condition particulière d'avoir résidé pendant au moins dix ans au Luxembourg et d'avoir préalablement participé aux élections communales ou européennes au Luxembourg?'

[60]TNS Ilres (2012).

referendum and the consequences it might have had.[61] It should nonetheless be noted that a prevailing argument in the debate leading up to the referendum was that if foreign residents can vote, Luxembourgers would not only soon be a minority in their own country, a minority in the decision-making body. Again, it is unlikely that in larger countries with a correspondingly different demographic composition, foreign residents could outweigh nationals in the electoral process. We may thus conclude that in light of Luxembourg's small 'native' population it has a particularly diverse population, a fact which allowed it to claim different legal status in EU law regarding the practical effects of EU citizenship. The next section explores a different area that shows that EU law indeed sometimes produces different effects in very small Member States.

4.4 The *Giersch* Case

This section examines a judgement of the CJEU that is probably the most recent well-known case involving Luxembourg. *Elodie Giersch v Etat du Grand Duché du Luxembourg* raised the question whether the children of cross-border workers in Luxembourg have the same right to funding for higher education as children of workers residing in the Grand Duchy.[62]

A little background to the legal question must be provided first. It has already been seen that the Grand Duchy's workforce is predominantly made up of non-nationals, that is to say foreign residents but also those workers that work in Luxembourg and reside in one of its neighbouring countries. Cross-border workers (the *'transfrontaliers'*) make up 44% of the Luxembourg workforce.[63] Together with foreign residents cross-border workers make up 71% of the Luxembourg workforce.[64] Under the previously existing scheme of higher-education grants, only those students with permanent residency in Luxembourg were eligible for financial support. This policy was however challenged in front of Luxembourg Courts by a number of children of cross-border workers, claiming that it was incompatible with the free movement of workers principle under Article 45 TFEU. This argument was raised in front of domestic Courts and led to a reference for preliminary ruling in front of the CJEU. This constituted one of the rarer instances of Luxembourg Courts proceeding to refer questions to the Court as Luxembourg judges rely on the preliminary reference procedure rather modestly.[65] Between 1963 to October 2013 only 87 requests for preliminary ruling were

---

[61] As well as the paradox of Luxembourg continuing to apply a restrictive regime with regard to local elections yet considering to have a way more inclusive regime regarding national elections than any other Member State at this moment in time.

[62] C-20/12 (2013) ECR 411.

[63] Eurostat (2016).

[64] There are also 10,000 people employed by the EU institutions in Luxembourg. See Government of the Grand Duchy of Luxembourg (2015).

[65] Mak (2010), p. 316.

introduced.[66] Half of these references were made by the highest civil or administrative Courts and none has ever been made by the Constitutional Court.

The question addressed to the CJEU was whether the fact that a Luxembourg law that made funding for higher education conditional on residence in Luxembourg amounted to discrimination on the basis of nationality. According to the opinion of Advocate General Mengozzi, the law indirectly discriminated against domestic and foreign workers but could be justified as it aimed at increasing the share of Luxembourgers with a higher education degree, a necessary pre-condition to realise the Grand Duchy's plan to transform itself into a knowledge-based economy.[67]

The Court disagreed. It declared the Luxembourg law incompatible with EU law on the ground that it went beyond what was necessary to achieve its stated objective: to increase the share of Luxembourg residents with a higher education degree. Rather, the Court held, the measure was to be qualified as a social advantage that must be granted to migrant workers under the same conditions as to national workers. The residence criterion imposed by the Luxembourg law was consequently found to amount to indirect discrimination on grounds of nationality as it operated to the detriment of frontier workers while benefitting those of Luxembourg nationality.[68]

The CJEU acknowledged that a residence criterion could work as an appropriate mechanism to lift the number of Luxembourg residents with a university degree but found the criterion of prior residence (i.e. that the student must reside in the Grand Duchy before taking up her studies) unjustified as this should not be the sole element taken into account to examine the degree of attachment to the State.[69] Indeed, the Court found that the children of frontier workers that have worked in Luxembourg for a long period of time may be just as likely to join the Luxembourg workforce. Alternative solutions would be available such as for instance a system where the grant of the loan or its reimbursement would be conditional upon joining the Luxembourg labour market or a condition that the student's parents must have worked in Luxembourg for a number of years (this last condition was subsequently adopted in the context of a reformed policy).[70]

The question that remains is what Luxembourg's small territorial size has to do with the *Giersch* case. At first sight—nothing much. A second look however reveals the reasons why Luxembourg attempted to shield its funding system from non-residents: namely because allowing non-residents access to the scheme was likely to radically increase the number of applicants, and recipients, of the funds, thus constituting a significant financial burden.[71] It is very unlikely that in a larger

---

[66]Gerkrath (2014), p. 120; Weirich (1986), p. 982; Schockweiler (1998) quoting Jean-Paul Hoffman: 'Les questions préjudicielles posées par les juridictions luxembourgeoises à la CJCM'.

[67]Opinion of Advocate General Mengozzi of 7 February 2013 in C-20/12 *Elodie Giersch v. Etat du Grand Duché du Luxembourg* (2013) ECR 411.

[68]Ibid, para 46.

[69]Ibid, para 63.

[70]Ibid, para 79.

[71]Ibid, para 50. In front of the CJEU, the Luxembourg government in fact argued that 'because of budgetary constraints it cannot be more generous towards non-resident students without putting the entire system of financial aid into question'.

state with a larger population the obligation to respect EU law in this manner would dramatically increase the number of beneficiaries of a higher education funding scheme—potentially more than doubling the number of recipients. The impact of the free movement of workers principle thus has budgetary impacts of a different magnitude in Luxembourg by virtue of its smallness. The next section ponders the relation between Luxembourg's small size and its fundamental rights regime.

4.5   The Lack of Fundamental Rights Catalogue in Luxembourg

It has been briefly indicated above that Luxembourg has been going through a process of constitutional revision (the 'refonte constitutionelle') lately.[72] This section engages with one aspect of the proposed new Constitution, which still has to be adopted, namely its fundamental rights regime. It will be seem that the fundamental rights aspect of the Draft Constitution has been subject to considerable criticism by academics and even by the European Commission for Democracy through Law (which is more commonly known as 'the Venice Commission').[73] This section will explore Luxembourg's fundamental rights regime, outline the criticisms that have been voiced, and speculate that if it were not for Luxembourg's small nature, the drafters of the documents would have been more likely to give in to these criticisms.

The 1868 Luxembourg Constitution, which is still in force but has been extensively amended over time, is very much a patchwork constitution. This characteristic also applies to the fundamental rights regime it enshrines. As currently phrased, human rights protection in the Grand Duchy encompasses some of the rights it already enshrined in the nineteenth century, such as the right to secrecy of telegrams,[74] but also fourth generation rights[75] and everything in between. This on the one hand entails that coherence is no characteristic of the Luxembourg fundamental rights regime, but on the other also that only some of the rights typically found in twenty-first century texts made their way into the Grand Duchy's constitutional document. As a result, the protection of fundamental rights in Luxembourg is not so much anchored in its own Constitution but rather in EU law and the European Convention on Human Rights ('ECHR'). In fact, it 'can certainly be said with no exaggeration that in view of the age and incomplete nature of the

---

[72]Gerkrath (2013a), p. 457.

[73]The text of the Draft Constitution can be accessed online. Commission des Institutions et de la Révision constitutionnelle (2015) Texte cordonne propose par la commission des institutions et de la revision constitutionelle. Available at http://www.referendum.lu/Uploads/Nouvelle_Constitu tion/Doc/1_1_6030%20version%2030.06.15.pdf. Accessed 13 July 2016.

[74]The Constitution of Luxembourg 1868, art 28.

[75]Ibid, art 11.

constitutional catalogue, it is the primary source of fundamental rights in Luxembourg'.[76]

One of the debates surrounding constitutional revision in the Grand Duchy is whether, and if so how, the constitutional human rights regime should be reformed. One option would have been to include a genuine fundamental rights charter into the new Constitution that could have been largely based on the Charter of Fundamental Rights of the European Union ('the EU Charter') and the ECHR. The parallel incorporation of both rights regimes into the Luxembourg Constitution would have been facilitated by the at least partial overlap between these two documents. To the surprise, and disappointment, of seemingly all commentators this has not however been the chosen path. To the contrary, the fundamental rights regime will very much retain its patchwork character, composed on the one hand by explicitly mentioned constitutional fundamental rights, and on the other hand by those rights arising under EU and ECHR law, which are of course applicable in Luxembourg, but not explicitly mentioned in the Draft Constitution. I will first announce which rights made their way into the Draft Constitution and then outline why it is deplorable that they are as limited in number and as incoherent in nature.

Chapter Two of the Draft Constitution deals with fundamental rights. It opens with a broad statement that 'human dignity is inviolable'.[77] Chapter Two then consecutively enumerates a number of rights. I will only mention a few here. Examples include the right to physical and mental integrity,[78] the right to freedom of thought, belief and religion[79] and the right to a private life.[80] The Draft Constitution also enshrines the principle of freedom of speech and of the press[81] and freedom of association.[82] Article 16 of the Draft Constitution declares the principle of equality before the law,[83] equality between women and men[84] as well as a rather curiously formulated general prohibition of non-discrimination according to which an individual cannot be discriminated against by virtue of her 'situation or personal circumstances'.[85] Here, one clearly wonders why discrimination on the basis of sex finds explicit mention, but other well-established principles under EU and ECHR law such as the prohibition of discrimination on grounds of national or social origin, race, religion, disability, age and sexual orientation do not.[86]

---

[76]Gerkrath (2014), p. 121.

[77]The Draft Constitution of Luxembourg 2015, art 12 (supra note 74).

[78]Ibid, art 13.

[79]Ibid, art 14.

[80]Ibid, art 15.

[81]Ibid, art 23.

[82]Ibid, art 25.

[83]Ibid, art 16(1).

[84]Ibid, art 16(3).

[85]Ibid, art 16(2).

[86]Protocol 12 to the European Convention for the Protection of Human Rights and Fundamental Freedoms, art 1; Charter of Fundamental Rights of the European Union, art 21.

While the above indicates that there is somewhat of a rights catalogue in the Constitution, a number of rights that one would expect to find in the constitutional document are rather notoriously absent, most oddly the right to life, the precondition to any other right.[87] Indeed, what is striking is that a number of rights and provisions enshrined in EU law and the ECHR have not made their way into the Constitution, a surprising fact given that Luxembourg has been a founding member of both institutions and likes to pride itself with being a loyal servant thereof. While it is true that EU law as well as the law of the ECHR are binding on Luxembourg anyways, the question arises why a Constitution of a human rights-endorsing country that completely updates its constitution in the year 2015 didn't chose to adopt a more ambitious regime. This question is all the more urgent given that ECHR and EU law do not apply to all situations in Luxembourg. The drafters of the new document have indeed argued that there is no need for a comprehensive constitutional catalogue of rights as international Treaties, including those of the EU and the ECHR prevail over domestic law, including constitutional law.[88] Strictly speaking, however, the judgements of the European Court of Human Rights ('ECtHR') are only binding inter partes and the EU Charter only applies only to Member States when they are 'implementing' EU law.[89] There would accordingly have been much value in creating a more comprehensive and inclusive constitutional rights catalogue, governing all situations arising in the Grand Duchy. What is more is that the coexistence of rights at national, EU and ECHR level, the so-called 'multidimensionality'[90] of rights in Europe poses a challenge in knowing to these different regimes relate to each other. One could also have hoped that the Draft Constitution would address this issue.

The commentary on the planned reform is quasi-unanimous. All commentators have deplored that the new constitutional rights regime lacks in ambition. Judge Spielmann, the Luxembourg judge and President of the ECtHR labelled the fundamental rights aspect of the new constitution as 'reassuring but of low ambition'.[91] He argues that a 'modern constitution must contain a more or less extensive catalogue of rights, which are justiciable before the Constitutional Court, which is the guardian of these rights'.[92] It is in this context that he deplores that the currently-proposed catalogue of rights is limited and imprecise.[93] Legal scholars agree. According to Gerkrath, it 'seems appropriate to update more radically the

---

[87] It might be argued that this right is implicit in the prohibition of the death penalty, which is enshrined in the Constitution. This would still not, however, explain why the drafters refrained from explicitly affirming it, given its centrality.

[88] Gerkrath (2013b).

[89] The Constitution of Luxembourg 1868, art 51.

[90] Garlicki (2008), p. 511.

[91] Spielmann (2001).

[92] Ibid, p. 19.

[93] Ibid, p. 19.

wording of Chapter 2'.[94] Others have joined these criticisms, suggesting that a pro-homine principle could be a valuable alternative in light of the opposition to the establishment of a fully-fledged rights charter.[95] The creation of such a charter has indeed been refused on the ground that there would be no clear distinction between directly effective fundamental rights and freedoms and mere constitutional 'objectives' that do not directly create fundamental rights.[96]

According to the Venice Commission, the current revision phase should be an occasion to proceed to a general updating of rights to correspond to what a modern constitutional on rights should be.[97] Observing that a number of internationally-recognised rights have not made their way into the Luxembourg Constitution,[98] it invited Luxembourg to do so to avoid any ambiguity regarding the status of these rights in Luxembourg law.[99] The absence of such an action was said to lead to a lack of cohesion.[100] The Venice Commission also deplored that the Draft Constitution does not determine the limitations to the rights it enshrines, which is a task for the legislator.[101] This, it went on to argue, attributes a lot of trust to the legislature as the Constitution provides that laws should determine the restrictions likely to affect rights and liberties.[102]

It appears likely that in a larger country with a larger critical mass of human rights NGOs, experts, academics, activists and interested bystanders more pressure would have accumulated to create a more comprehensive human rights regime. Such pressure is however more difficult to gain momentum in a small country with a very small critical mass and no international attention to domestic constitutional events. Luxembourg, with its occasional shortcomings, has a good record of human rights protection and there does not appear to be any principled opposition towards the continued protection of such rights. In light of these facts it is easy to imagine that more pressure could quickly have led to a different approach towards the protection of fundamental rights in the new constitution, in line with the regime established under the law of the European Union.

---

[94]Gerkrath (2013a), p. 457.

[95]Bruck (2014).

[96]Gerkrath (2013a), 'Some Remarks on the Pending Constitutional Change in the Grand Duchy of Luxembourg', p. 456.

[97]The Venice Commission (2009), para 34.

[98]The Venice Commission (2009).

[99]Ibid, para 35.

[100]Ibid, para 36.

[101]Ibid, para 39.

[102]Ibid.

4.6   Conclusion: Territorial Size as One of the Factors Impacting on the Relation Between National and Supranational Legal Orders

This brief chapter has examined a number of aspects of the relationship between the law of the European Union and Luxembourg law. This has included a survey of the relation between Luxembourg law and the EU doctrine of supremacy, the reform of voting rights as a consequence of European integration, the impact of the EU principle of free movement of workers on the budget of a small State, and fundamental rights reforms. I have suggested that in each of these scenarios, smallness constitutes a useful lens that allows us to understand the relationship between a small state and the supranational legal order.

While putting forward that Luxembourg's small size is an important factor concerning the relation between EU law and domestic (constitutional) law, I am not claiming that territorial or demographic size is the sole factor impacting on the relation between a domestic legal order and European Union law. Rather, I suggest that it is one of many. While we already know that other important factors in this regard are particular constitutional traditions or political influences, smallness has not yet been explored in this context. This chapter is but a small indication that territorial and demographic size might matter in this regard and that this is a topic that deserves further exploration in subsequent research.

The aim and results of this chapter were not to scientifically establish the exact impact smallness has on these aspects, which without a doubt are influenced by a number of factors. Nonetheless the results of this research have shown that size may indeed matter when it comes to the relation between small States and the legal orders they are embedded in. Smallness may in many aspects mean sensitivity: certain legal provisions can have more pronounced effects in smaller states. This has been underlined in this chapter through the survey of how principles of EU free movement have different budgetary consequences in Luxembourg than they would in larger States. In 1974, Lord Denning said in *Bulmer v Bollinger* that EU law is 'like an incoming tide. It flows into the estuaries and up the rivers'.[103] Just as rivers, Member States are however of different sizes, and just as tides have different effects on rivers depending on whether they are large or small, a different image emerges with regard to the relation between EU law and smaller or larger Member States in some respects. This is one of the many factors worthy of bearing in mind when attempting to make sense of the relation between EU law and domestic legal orders.

---

[103][1974] EWCA Civ 14, [1974] 2 All ER 1226, [1974] 3 WLR 202, [1974] Ch 401.

References

Bruck V (2014) Mieux proclamer pour moins protéger? critique de dévalorisation des droits de l'homme par le projet de constitution. Forum 339:7–9

Craig P, de Búrca G (2011) EU law: text, cases, and materials, 6th edn. Oxford University Press, Oxford

de Witte B (2011) Direct effect, primacy, and the nature of the legal order. In: Craig P, de Búrca G (eds) The evolution of EU law. Oxford University Press, Oxford

Der Internationale Karlspreis zu Aachen (2016a) Le prix international Charlemagne d'aix-la-chapelle. Available at http://www.europarl.europa.eu/summits/edinburgh/a6_en.pdf. Accessed 7 July 2016

Der Internationale Karlspreis zu Aachen (2016b) Charlemagne prize laureate 1986: the people of Luxembourg. Available at http://www.europarl.europa.eu/summits/edinburgh/a6_en.pdf. Accessed 7 July 2016

Dicey AV (1915) Introduction to the study of the law of the constitution, 8th edn. Macmillan Publishers, London

European Council on Edinburgh (1992) Conclusions of the presidency, annex 6, part A. SN465/1/92 Rev 1. Available at http://www.europarl.europa.eu/summits/edinburgh/a6_en.pdf. Accessed 13 July 2016

European Justice (2016) Member state law, Luxembourg: sources of law. Available at. https://e-justice.europa.eu/content_member_state_law-6-lu-en.do Accessed 13 July 2016

European Union (2016) Luxembourg. Available at http://europa.eu/about-eu/countries/member-countries/luxembourg/index_en.htm. Accessed 7 July 2016

Eurostat (2016) Migration and migrant population statistics. Available at http://ec.europa.eu/eurostat/statistics-explained/index.php/Migration_and_migrant_population_statistics. Accessed 13 July 2016

Finck M (2015) Towards an ever closer union between residents and citizens? On the possible extension of voting rights to foreign residents in Luxembourg. Eur Const Law Rev 11(1):78–98

Garlicki L (2008) Cooperation of courts: the role of supranational jurisdictions in Europe. Int J Const Law 6(3):509–530

Gerkrath J (2013a) Some remarks on the pending constitutional change in the grand duchy of Luxembourg. Eur Public Law 19(3):449–459

Gerkrath J (2013b) Constitutional amendment in Luxembourg. In: Contiades X (ed) Engineering constitutional change: a comparative perspective on Europe, Canada and the USA. Routledge, London

Gerkrath J (2014) The figure of constitutional law of the 'integrated state': the case of the grand duchy of Luxembourg. Eur Const Law Rev 10(1):109–125

Gerkrath J, Thill J (2012) The grand duchy of Luxembourg. In: Besselink L, Bovend'Eert P et al (eds) Constitutional law of the EU member states. Kluwer, The Netherlands, pp 1085–1145

Government of the Grand Duchy of Luxembourg (2015) Luxembourg job market. Available at http://www.luxembourg.public.lu/en/travailler/marche-emploi/index.html. Accessed 10 July 2016

Heuschling L (2013) La citoyenneté de résidence, diverses logiques et la science juridique. Forum 326:32–37

Index Mundi (2015) Luxembourg demographics profile 2014. Available at http://www.indexmundi.com/luxembourg/demographics_profile.html. Accessed 10 July 2016

Institut national de la statistique et des études économiques (2015) Communiqué de presse: informations statistiques recentes. Available at http://www.statistiques.public.lu/fr/actualites/population/population/2015/04/20150408/20150408.pdf. Accessed 10 July 2016

Kinsch P (2010) Le rôle du droit international dans l'ordre juridique luxembourgeois. Pasicrisie Luxembourgeoise 34

Kumm M (2005) The jurisprudence of constitutional conflict: constitutional supremacy in Europe before and after the constitutional treaty. Eur Law J 11(3):262–307

Lansbergen A, Shaw J (2010) National membership models in Europe. I-CON 8(1):50–71
Lardy H (1997) Citizenship and the right to vote. Oxf J Legal Stud 17(1):75–100
Mak E (2010) Report on the Netherlands and Luxembourg. In: Martinico G, Policing O (eds) The national judicial treatment of the ECHR and EU laws: a comparative constitutional perspective. Europa Law Publishing, Groningen
Moyse F (2005) La Charte européenne des droits fondamentaux et son application en droit luxembourgeois: une avancée modérée pour les droits de l'homme. Available at http://www.forum.lu/referendum/article7597.html?id_article=320. Accessed 10 July 2016
Pauly M (2013) Histoire du Luxembourg. Université de Bruxelles, Brussels
Pescatore P (1962) L'autorité en droit interne, des traites internationaux. Pasicrisie Luxembourgeoise
Ravarani G (2007) La Constitution luxembourgeoise au fil du temps. In: Annales du droit luxembourgeois, vol 17. Bruylant Publishers, Belgium, pp 1817–1818
Schockweiler MF (1998) Bulletin du cercle Francois Laurent. Luxembourg, pp 5–79
Spielmann D (2001) Luxembourg. In: Blackburn R, Polakiewicz J (eds) Fundamental rights in Europe: the European convention on human rights and its members states. Oxford University Press, Oxford, pp 1950–2000
The Venice Commission (Commission Europeenne pour la Democratie par le Droit) (2009) Avis interimaire sure le project de revision constituionelle du Luxembourg. CDL-AD(2009)057: 544/2009. Available at http://www.venice.coe.int/webforms/documents/default.aspx?pdffile=CDL-AD(2009)057-f. Accessed 13 July 2016
TNS Ilres (2012) Sondage TNS ilres: partie 4, droit de vote. Available at http://www.2030.lu/fileadmin/user_upload/documents/2030.lu_TNS_ILRES_Droit_de_vote_FR_28032013.pdf. Accessed 10 July 2016
Trausch G (1992) Histoire du Luxembourg. Hater Publishing, Paris
von Bogdandy A, Schill S (2011) Overcoming absolute primacy: respect for national identity under the Lisbon treaty. Common Mark Law Rev 48(5):1417–1453
Weber P (1957) Les constitutions du 19e siècle. Livre jubilaire du conseil d'etat 4(4):303–362
Weirich M (1986) L'application du droit communautaire au Grand-Duché de Luxembourg. In: Livre jubilaire de la Conférence Saint-Yves. Saint Paul, Luxembourg, pp 965–1004

Chapter 5
The Taxation of Small States and the Challenge of Commonality

Ann Mumford

5.1 Introduction

The starting point for this chapter is to consider small states, and issues that might be specific to them, in the context of international tax law. At the moment, international tax law is a remarkably energised subject, with a great deal of discussion focused on initiatives to tackle challenges surrounding multinational corporations, in particular. The topic of this collection presumes a commonality of interests amongst small states; thus, this chapter seeks to investigate whether this commonality extends to international tax law. As the analysis which follows will seek to demonstrate, this topic raises many questions about the nature of transnational consensus, and of transnational law in general.

To begin: what is the value of considering taxation in the context of the size of the state? It is important to identify the purpose of the exercise. It may be that the object of the exercise is to demonstrate the comparative lack of influence of smaller states in international discourse on taxation initiatives. Alternatively, it may be that the intention is to demonstrate that international initiatives, or transnational movements, impact disproportionately on smaller states than on larger states. Finally, it is possible that the exercise is without preconception, and, rather, an exercise of strategy. By asserting a commonality of interest, smaller states, collectively, may hope to 'punch above their weight' in the international tax sphere, and to influence negotiations to the same extent as larger states.

This chapter aims to explore the value of the context of size, and to analyse the challenge of commonality, in the consideration of international taxation issues as they pertain to small states. It addresses four, specific issues. First, is size an important factor in the negotiation process for bilateral taxation treaties? Are

A. Mumford (✉)
Dickson Poon School of Law, King's College London, London, UK
e-mail: ann.mumford@kcl.ac.uk

© Springer International Publishing AG 2017
P. Butler, C. Morris (eds.), *Small States in a Legal World*,
The World of Small States 1, DOI 10.1007/978-3-319-39366-7_5

smaller states less likely to achieve an outcome that is advantageous to them? Second, this chapter asks whether it is possible to locate questions pertaining to size and treaty negotiations within the context of transnational discourse. Third, literature pertaining to transnational consensus is reviewed, with particular attention paid to the role of size for (alleged) participants. Finally, this chapter considers the role of vulnerability in the transnational discourse of consensus, again, with particular attention paid to size.

5.2   Is Size an Important Factor in the Negotiation Process for Bilateral Taxation Treaties?

Size may appear less relevant when one considers that the starting point of international taxation is the bilateral tax treaty.[1] Two parties, thus, negotiate the terms of the treaty—so, perhaps, the balance is 50/50? Of course not, one may presume—one party is very likely to be more powerful than the other.[2] The lack of equality in bargaining position, however, need not necessarily result from comparative size— though equally, of course, it may. There is an additional layer of complexity in that, as Avery Jones famously observed, the treaty negotiation process does not encourage participants to represent themselves accurately. Each party will be encouraged to emphasise the potentially harsh consequences for the other country's taxpayers of the domestic position, in order to encourage concessions in the negotiations.[3] The reflections of a country's true tax system that emerge from bilateral tax treaty negotiations need not be accurate—and, indeed, may be distorted. In this, however, taxation is not different from any other aspect of international law,[4] and, as the other chapters in this collection demonstrate, small countries may possess a collective sense of vulnerability[5] in other international contexts as well.

Partly because of these and other uncertainties, the bilateral tax treaty has been criticised as an ineffective tool for the twenty-first century, multilateral world.[6]

---

[1]Friedlander and Wilkie (2006), p. 909: The origin of modern tax treaties has been traced to nineteenth century "'friendship, commerce and navigation" treaties'.

[2]Elfman (1995), p. 177: Addressing the topic of vulnerability (the concept of vulnerability is discussed in greater detail in this chapter), and summarising the position of Walt (1987), pp. 21–31, Elfman writes '[s]ince weak states are vulnerable to the aggressive demands of great powers, they will ally with a dominant power in order to avoid immediate attack'. Weak is equivalent to 'small' in this observation.

[3]Avery Jones (2000), p. 3: '[t]he more outrageous the provisions of internal law, the better the starting position for negotiating treaties'.

[4]Also note Avi-Yonah (2004), p. 483: 'Is international tax law part of international law? To an international lawyer, the question posed probably seems ridiculous... [but] once one delves into the details, it becomes clear that in some ways international tax law is different...'.

[5]McLean (1985); Charles (1997).

[6]Avery Jones (2000), p. 3: '[t]he disadvantage of the tax treaty route is that it is self-perpetuating. Treaties are a one-way street; they lead only to more treaties'.

From the 1920s through the 1960s, double taxation was the sole target of the transnational legal order underpinning international taxation, resulting in a transnational legal order with 'a high level of issue alignment'.[7] There were several unintended consequences of this, including the encouragement of conditions which would help tax competition, whilst simultaneously restraining the capacity of governments to control or to dissuade tax competition.[8]

Multilateral agreements would be preferable to the current state of affairs, the argument continues, but have difficulty emerging from the traditional negotiating process. Yet even as the OECD and the UN have supported the bilateral treaty process through their model tax treaties, they also have spearheaded initiatives which are inherently multilateral. The OECD's Harmful Tax Competition project,[9] and the current Base Erosion and Profit Shifting programme,[10] are predicated on assumptions of transnational consensus. The essential concept of transnational consensus may appear to be a twenty-first century idea, but in many ways it is just a modern iteration of the League of Nations' belief that global resources could only be spread more equitably through the proliferation of global trade; and, thus, it is important to remove the barriers to trade posed by taxation.[11] Double taxation, it was argued, would encourage wealth to stay at home. Remove this threat, and wealth would travel.

What perhaps was not anticipated was that wealth, indeed, would travel, yet in some instances it would travel because of tax, or lack of it. Trade is not the only engine of the global marketplace. Perhaps the distinction—between motivation based on taxation, or trade—always has been academic, and with diminishing practical relevance. Businesses may be willing to travel for profit—whether that profit originates from 'true' economic activity, or from tax efficiency, may be of interest to the tax authority, or a government, but of less interest to the business

---

[7]Genschel and Rixen (2015), p. 155.

[8]Ibid.

[9]Guttentag (2001), pp. 548–549.

[10]Baker (2015), pp. 85–86: Discussing the history of the OECD's development of the phrase 'aggressive tax planning', beginning with 2002; Panay (2016): The introduction of BEPS is described as '[a] realignment of taxation and relevant substance was, therefore, required, as international tax standards had not kept pace with changing business models and technological developments'.

[11]Guttentag (2001), pp. 549–550: Discussing the modern challenges of bilateral tax treaties; Avery Jones et al. (2006): Considering the history of the OECD Model Treaty, Avery Jones et al. explain that the OECD Model 'developed out of the League of Nations Models which were strongly influenced by the treaty practice between the mainland European countries. For that reason, a common law reader coming to the OECD Model for the first time might find it full of unfamiliar expressions'. This article presents the results of a survey of several countries, and their connection to terms and expressions found in the OECD Model tax treaty. It concludes that many of these expressions can be traced to those used in civil law European countries just after the First World War.

person. When the question of size is considered against the background of these concerns, thus, it does appear relevant, as size is connected to wealth and dominance. The suggestion is that the modern international tax system is not as capable as it might be of compensating for the disadvantage of size.

5.3 Is It Possible to Locate Questions Pertaining to Size and Treaty Negotiations Within the Context of Transnational Discourse?

The relevance of size perhaps increases within the concept of a *transnational* consensus. A transnational legal order (TLO) aligned to prevent double taxation in part produced tax competition, and efforts to redress this may impinge on national sovereignty in ways that governments may find unappealing.[12] Small states may not have contributed to the norms underpinning the transnational legal order, but nonetheless are impacted by it.

Transnational law describes the collection of practices, rules and customs that transcend domestic legal systems, and appear to govern what would be understood within tax legal discourse as multilateral problems. The literature relating to double taxation treaties highlights the difficulties of multilateral problems,[13] but does not ask whether transnational responses, or consensus, fill the gaps that have been left by the bilateral treaty.

The challenge is to identify the difference between the concepts of multilateral, and transnational. At first glance, it might appear to be straightforward—multilateral involves the agreement of several, traditional legal systems, whereas transnational deals more with governance that arises from non-traditional sources. The difficulty with this assumption is that it is based on the suggestion that the definition of transnational is generally accepted, and even a brief review of the literature reveals that it is not. Cotterrell suggested that European law inhabits the transnational sphere, simply because it 'spills out' from the borders of the nation state.[14] So, the definition of transnational is clearly somewhat fluid; or, at the least, not rigid. He explains that this relatively new term, transnational law, has seemed necessary to indicate new legal relations, influences, controls, regimes, doctrines, and systems that are not those of nation-state (municipal) law, but, equally, are not fully grasped by extended definitions of the scope of international law'.[15] Although he concedes that the term is not used with much 'precision',[16] the 'transnationalisation' of European law in particular is perhaps best understood not

---

[12]Genschel and Rixen (2015), pp. 157–158.

[13]Calliess and Zumbansen (2010), p. 6.

[14]Cotterrell (2012), p. 500.

[15]Ibid, pp. 500–501.

[16]Ibid, p. 501.

as an end product of the integration and harmonisation of the different legal systems of the member states, but, rather, as the process by which the policymaking agenda of the EU is set.[17] In other words, policymakers within the EU view themselves as working within a wider, transnational context. Yet, the perspectives of small states have been noticeably absent from the literature analysing the role of the state or the 'municipal' in European law; and, indeed, the combined and individual influence of France, Germany and the United Kingdom appears to drive this area of scholarship.[18]

Additionally, the lack of precision in terms is not without its difficulties. Avi-Yonah, addressing this subject broadly but speaking of 'multinational enterprises' specifically, warns that '[t]he choice of terminology in this field is inevitably value-laden'.[19] As he explains, 'MNEs [MultiNational Enterprises] is the preferred tern of the rich countries and the OECD; developing countries and the UN prefer to call them transnational corporations (TNCs)'.[20] Interestingly (yet also within the specific context of his article), Avi-Yonah explains that he prefers to use the term MNEs, because TNCs 'are typically not one corporation'.[21] Thus, he explicitly rejects the option of employing the term transnational, as not appropriate to his analysis of the taxation and regulation of multinational corporations. It would appear that there is indeed a place for consideration of the question of the taxation of small states within discussions of transnational law—the challenge, however, is to identify the normative values underpinning these discussions, and to avoid the distraction of historical processes.

5.4 Consideration of the Literature Pertaining to Transnational Consensus, with Particular Attention Paid to the Role of Size for (Alleged) Participants

The latter statement is perhaps best understood within the context of what constitutes an historical 'consensus'. There is a wider literature on transnational and multilateral consensus, with particular relevance for questions of size. Cotterrell challenges the concept of multilateral, perhaps understood instinctively in tax as, simply, more than two. He approaches the question from the point of consensus, or by comparing a unilateral agreement to a multilateral agreement.[22] With a multi-lateral agreement, a group has decided something, but then it will be necessary to put in place '(Hartian) secondary rules' to enforce the agreement, or at least an

---

[17]Kaiser and Starie (2005), p. 2.

[18]Half and Soetendorp (1998), pp. 3–4.

[19]Avi-Yonah (2003), p. 5, fn. 1.

[20]Ibid.

[21]Ibid.

[22]Cotterrell (2012), p. 507.

agency to oversee its execution.[23] Similarly, Calliess and Zumbansen addressed the concept of consensus, suggesting, first, that they 'understand transnational law above all to demarcate a methodological position rather than to identify a perfectly map-able doctrinal field'.[24] They endeavour thus to engage with a space which is 'captured' neither by public nor private law, whilst demonstrating that boundaries between both categories of law, increasingly, are less relevant.[25] The idea of addressing small states, and their taxation, would appear to emanate from a denial of consensus, and perhaps more in line with analyses offered by scholars such as Danielsen, who has called for greater attention to be paid to the extent to which corporations influence what he describes as 'transnational regulation', using many of the same illustrations as Cotterrell to populate this area of law (essentially, anything transcending the domestic, or, as Cotterrell would describe, the 'municipal').[26]

Avi Yonah (whilst not using the term transnational) suggests that the taxation of multinational enterprises requires 'unilateral extraterritoriality with reciprocity'.[27] It is instructive to consider this definition within the context of perhaps one of the most famous examples of small countries entering into consensus, and the challenges this posed for definitions of the 'Washington Consensus'. The concept of the 'Washington Consensus' of the early 1990s, roughly, is used to describe the process within which developing countries '. . .privatized state-owned industries, removed trade barriers and generally moved towards increased reliance on state intervention in their economies'.[28] Babb suggests that the Washington Consensus is a 'transnational policy paradigm', largely because of its unique origins.[29] She argues that economic scholarship was relied upon to give this process an aura of legitimacy, whilst both domestic governments and international organisations[30] collaborated and 'encouraged' the process.[31] This 'paradigm' policy moment, however, was a brief historical moment, ultimately 'weakened' by '. . .its own internal vulnerabilities and the changing intellectual and political circumstances'.[32] It has not yet been completely replaced, however, as nothing strong enough has yet emerged to replace it.[33]

Returning to Avi Yonah's definition—'unilateral extraterritoriality with reciprocity'—it is important to remember that he was describing the process of taxing

---

[23]Cotterrell (2012), p. 507.

[24]Calliess and Zumbansen (2010), p. 6 (supra note 13).

[25]Ibid.

[26]See for example Danielsen (2005). See also, generally, Likosky (2002).

[27]Avi-Yonah (2003), p. 32.

[28]Baby (2013), p. 268.

[29]Ibid.

[30]Ibid: What she describes as 'international financial institutions'.

[31]Ibid.

[32]Ibid, p. 289.

[33]Ibid.

multinational corporations, or, rather, the justification by which states do so. By 2016, the literature has evolved towards considering the taxation of multinationals from a variety of perspectives; and, in particular, the concept of 'detaxation' has emerged. The process of detaxation[34] depletes tax from the state, generally in favour of the stateless, multinational corporation. Domestic tax laws within this context may be viewed from two perspectives. The first is that domestic tax laws increasingly are meaningless in the face of transnational initiatives and globalised economic mobility. The suggestion is that the League of Nations' unintended legacy is that by forging a system in which some tax laws may be discounted through a bilateral agreement, a process was created in which all tax laws might potentially be avoided. The second perspective, however, is that domestic tax laws are increasingly more important; in some ways, infinitely more valuable. They resist the transnational pull to detaxation, and persist in funding the increasingly vulnerable state.

The question which follows, however, within this chapter, is the *size* of the state. Indeed the conceptual challenge for this chapter is the small state which has increased its ability to compete internationally, and indeed to attract the share of global wealth that the targeting of double taxation was intended to achieve, by offering its legal system as a vehicle, in effect, for detaxation. So, global resources have been redistributed—in some ways, this is exactly the outcome for which the League of Nations had been hoping.[35] The criticism lies in the pattern of the redistribution, for the suggestion is that far too much benefit has accrued to the elusive taxpayer company, and far too little to states of any size.

The goal of the League of Nations was not simply to spread the world's wealth more equitably, but to do so through trade, and trade is missing from the pattern described above. The small state has not 'traded' internationally, but, rather, served as a host, or, more critically, an accomplice. The small state has not contributed to the sum total of global economic activity, though it has played a role in diminishing the resources available to fund another (large?) state, which was a consequence of the League of Nations' project. Corporations were valuable in this early vision only insofar as they contributed to taxation, and whilst they may be contributing (whether through employment or a smaller amount of taxation) to the overall wealth of a small state, the suggestion is that the proportion of the contribution is far too restricted. Indeed, Sharman has suggested that small states have 'nothing to lose through continued recalcitrance' towards OECD initiatives, largely because it is increasingly impossible for them to meet the standards of existing programmes (referring to the Harmful Tax Competition (HTC) project).[36]

---

[34]Lahey (2015): The concept of detaxation as discussed throughout this chapter is based upon the definition developed by Lahey in this article.

[35]Coates (1924). See also Avery Jones (2013): 'One would have liked to have been able to say that the 1923 Report was the economic foundation for the future of double taxation relief, or even of tax treaties, but that is not the case. . .'.

[36]Vlcek (2008), p. 4 citing Sharman (2005), p. 317.

Yet what Avi Yonah describes as the 'current age of globalization' differs in a number of respects from other historical periods in that capital is, in fact, much more mobile, especially as compared with labour.[37] He explains that the challenge of consensus in tax is that often the policies of different countries will complement each other, but multinational enterprises will be keen to prevent these policies from applying 'throughout the enterprise'.[38] Thus, '[c]ountries are in general agreement that the profits of MNEs should only be taxed once, but the MNEs, while seeking to prevent double taxation, do not object to double non-taxation'.[39] Identification of a 'consensus', thus, appears to risk allocation of blame, possibly even directing blame (illogically) towards smaller, patently more vulnerable states.

5.5 The Role of Vulnerability in the Transnational Discourse of Consensus, with Particular Attention Paid to Size

As Vlcek presciently observed in 2009, three themes come to mind when addressing the challenge of taxing small states: sovereignty, size and money. The transposition of taxation into the transnational sphere has the potential to increase the vulnerability of small states. Indeed Charles has explained that overcoming the sense of vulnerability in taxation is one of the significant challenges for small states. The vulnerability, she explains, emanates from a general sense of overexposure to risks which larger states are simply able to absorb.[40] 'Serious environmental risks' are an example of such a vulnerability[41]—perhaps understandable given that many small states are also coastal states, though of course not all small states are (and, indeed, the unique vulnerabilities of small states to climate change seem less unique as environmental disasters proliferate in coastal areas globally, both in small states and large). The absorption concept is perhaps especially important in the context of risks posed to small states by decisions which appear to have been taken by larger states. Thus, is the international tax system truly a product of an (albeit, 'rough', in the sense analysed by Callieff and Zumbansen) consensus; or—and it is in perhaps this sense that Sharman's description of 'recalcitrance' starts to take hold—is it the product of decisions taken either by larger states, or their corporations?

The distinction between the larger states, and corporations, is important, because it speaks to potential remedy, as one response to the detaxation trend is to call for the strengthening of the state and its domestic tax rules. Indeed, one response to transnational law, generally, is to decry it as capitalism supplanting

---

[37]Avi-Yonah (2001), p. 61.

[38]Avi-Yonah (2003), p. 11.

[39]Ibid.

[40]Ibid.

[41]Ibid.

the state[42]—and, in this sense, *re*taxation would provide a retaliation. The challenge is to find a place for small states in a retaxation process in which their vulnerabilities are addressed, and their fear of overexposure to risk is considered.

The concept of vulnerability is a unifying thread in much of the literature pertaining to small states. Charles explains that these vulnerabilities extend to worries over insularity (are small states sufficiently outward looking?), although insularity perhaps may be mollified if perceived, rather, as social cohesion. Large states are not by definition less insular than small states—and, indeed, the classical system of corporation tax in the United States is sometimes explained by an economy which historically has been more concerned with internal trade, than international trade[43]—and, indeed the drive to classification or description of small states may appear to falter at ontology, in that there are patent dangers of generalisation.

One question is whether a process of *re*taxation would necessarily demand that smaller states isolate themselves, and retreat into the 'vulnerability' of insularity. As Koh observed when considering the transnational project, '[g]enerally speaking, the transnationalists tend to emphasise the interdependence between the United States and the rest of the world, while the nationalists tend to focus more on preserving American autonomy'.[44] This observation, on its own, would appear to highlight the significance of size, for if the transnational project is viewed from the perspective of the United States, the quintessential large, wealthy state, then it is a potential force for global engagement and distribution of wealth—and, yet, when viewed from the perspective of a smaller state, then the transnational project carries the vulnerabilities associated with joining a project the terms of which it did not dictate, from a perspective of comparatively reduced standing and influence. Additionally, from the perspective of the smaller state, the project may not be described accurately as *re*taxation; but, rather, as the establishment of a tax system *de novo*, on terms which it would not have chosen.

This is the predominant impression of small states in the context of transnational taxation policy—'a large number of relatively small places functioning as offshore financial centres'.[45] In a provocatively titled essay, Hampton and Christensen suggested that such states might be labelled 'offshore pariahs', largely as a consequence of the Harmful Tax Competition project started by the OECD in 1998.[46] The authors explain that '[d]espite extensive lobbying', only six of

---

[42]Robinson (2004).

[43]'The modern [US] corporation fits awkwardly into a set of tax principles based on economic and political theories that are drawn largely from a simplified picture of a society in which production is organized by small-scale proprietorships and partnerships...' per Goode (1951) cited in Bank (2010), p. ix; But compare Harris (2013), s 1.1.3 observing that '[i]n an increasingly globalized world, countries are now commonly faced with a multitude of corporate laws, and there is much to be learned from the US approach in this regard'.

[44]Koh (2005), p. 749.

[45]Cornell Cobb (1998), p. 7.

[46]Hampton and Christensen (2002), p. 1660.

forty-one jurisdictions succeeded in removing their countries from the Fiscal Affairs Committee initial list of 'tax havens'.[47] In a useful review of the literature, they highlight research which suggests that the 'relatively few' economically successful small states had succeeded specifically because they 'actively manage their dependency on large countries'.[48]

And, thus, this chapter returns to questions posed at its outset. A review of the literature does not reveal size to be a significant issue in the bilateral treaty negotiation process. Size does appear to be significant, however, when it comes to setting the agenda. Transnational consensus are reached—different OECD initiatives on tax abuse are examples of such agreements—but the participation of smaller countries in such agreements appears to be suspect. This chapter also sought to locate questions pertaining to size within the context of transnational discourse, and that this issue was approached from two perspectives—from the question of vulnerability (i.e., if the transnational consensus is that tax abuse should be targeted, then it is likely that smaller countries will be vulnerable within this), but also from the question of complexity. On this, the chapter posed more questions that it answered, especially from the perspective of the challenges posed by the multinational corporation. It would seem that the multinational corporation is the ultimate challenge to, and product of, transnational discourse.

5.6   Conclusion

And yet ultimately this chapter, in its examination of the 'challenge of commonality', sought to consider whether anything was to be gained from considering the question of taxation from the perspective of the size of countries. At the outset there are several reasons that suggest that the answer is no, not the least because there would appear to be better alternatives for analysis. Taxation might be considered, for example, from the perspective of developing countries, the perspective of the global south (or north), or, the perspective of non-OECD member countries. Resisting the perspective of size, however, also would result in resisting a classification with a rich history (consider the Commonwealth Secretariat's report), and which overlaps with several of the categories listed as alternatives (as small states often do have developing economies, for example). In particular, resisting the relevance of size would perhaps render the consideration of the role of transnational consensus less rich. This is because, as this chapter argued, the literature relating to double taxation treaties highlights the challenges posed to bilateral treaties by multilateral problems, although it does not ask whether transnational responses, or consensus, fill the gaps left by the bilateral treaty.

---

[47]Hampton and Christensen (2002), p. 1660.

[48]Ibid, p. 1663 citing Baldacchino (1993).

Again, however, questions were not settled, and were more frequently posed—perhaps this is not surprising, given that the definition of transnational, as this chapter discussed, is not agreed. The approach suggested by Calliess and Zumbansen—i.e., that they 'understand transnational law above all to demarcate a methodological position rather than to identify a perfectly map-able doctrinal position'—is particularly helpful to the question of tax, as it starts from a position of identifying problems for which existing legal orders may not have provided answers. In some ways, it is anthropological in potential, for it starts at the point of the problem—in this chapter, the problem would be the juxtaposition of the tax status of the multinational company (in Avi Yonah's description) with the sense of vulnerability of small states—and then awaits, or traces the emergence, of legal responses.

The possibility of overexposure to risks from which larger states may have more resilience is significant, as this chapter has argued, and may threaten ultimately the delicate consensus as to the tax status of multinational corporations. That consensus, in an era in which detaxation is identified as a major threat to the economic and social well being of states of all sizes, is that multinational corporations should pay more tax, and that transnational agreements should be forged to achieve this. This chapter does not wish to suggest that this is a consensus in which all small states have participated; and, in fact, it would rather argue that a transnational case for the *re*taxation of states, and from multinational corporations in particular, has emerged. There are two problems: double taxation, and tax competition.[49] The former hinders international trade, and produces 'welfare-reducing economic under-integration'; the latter, trade that is inefficient, risks 'welfare-reducing over-integration'.[50] 'The tax challenges posed by multinational companies in both of these problems have not been sufficiently addressed by existing national regimes (nor especially by bilateral tax treaties), and thus, in this space, one must wait to see what emerges. The role of international organisations such as the OECD clearly will be significant in this. But it would be wrong to assume that the role played by small states will not be significant as well—at the least because the lesson of history is that the risks of new global initiatives will be borne by them, perhaps most dearly, in the midst of an agenda of which they may not feel most fully in control, and which may not most explicitly be targeted to their benefit.

References

Avery Jones JF (2000) Are tax treaties necessary? Tax Law Rev 53:1–39
Avery Jones JF (2013) Sir Josiah Stamp and double income tax. In: Tiley J (ed) Studies in the history of tax law, vol 6, chap 1. Hart Publishing, Oxford

---

[49]Genschel and Rixen (2015), p. 154.
[50]Ibid.

Avery Jones JF et al (2006) The origins of concepts and expressions used in the OECD model and their adoption by states. Bull Int Tax 60(6):220

Avi-Yonah RS (2001) Globalisation and tax competition: implications for developing countries. Law Quad Notes 44(2):60–65

Avi-Yonah RS (2003) National regulation of multinational enterprises: and essay on comity, extraterritoriality, and harmonization. Columb J Transnatl Law 42(5):5–34

Avi-Yonah RS (2004) International tax as international law. Tax Law Rev 57(4):483–501

Baby S (2013) The Washington consensus as transnational policy paradigm: its origins, trajectory and likely successor. Rev Int Econ 20(2):268–297

Baker P (2015) The BEPS project: disclosure of aggressive tax planning schemes. Intertax 43:85–90

Baldacchino G (1993) Bursting the bubble: the pseudo-development strategies of microstates. Dev Change 24(1):29–51

Bank SW (2010) From sword to shield: the transformation of the corporate income tax, 1861 to present. Oxford University Press, Oxford

Calliess GP, Zumbansen P (2010) Rough consensus and running code: a theory of transnational private law. Bloomsbury Publishing, New York

Charles ME (1997) A future for small states: overcoming vulnerability. Commonwealth Secretariat, London

Coates WH (1924) League of nations report on double taxation submitted to the financial committee by professors Bruins, Einaudi, Seligman, and Sir Josiah Stamp. J Roy Statist Soc 87(1):99–102

Cornell Cobb S (1998) Global finance and the growth of offshore financial centres: the manx experience. Geoforum 19(1):7–21

Cotterrell R (2012) What is transnational law? Law Soc Inq 37(2):500–524

Danielsen D (2005) How corporations govern: taking corporate power seriously in transnational regulation and governance. Harv Int Law Rev 46:411–425

Elfman MF (1995) The foreign politics of smalls states: challenging neorealism in its own back yard. Br J Polit Sci 25(2):171–217

Friedlander L, Wilkie S (2006) Policy forum: the history of tax treaty provisions - and why it is important to know about it. Can Tax J 54(4):907–921

Genschel P, Rixen T (2015) Settling and unsettling the transnational legal order of international taxation. In: Halliday TC, Shaffer G (eds) Transnational legal orders. Cambridge University Press, London

Goode R (1951) The corporation income tax. Wiley, New York

Guttentag JH (2001) Key issues and options in international taxation: taxation in an interdependent world. Bull Int Tax 55(11):546–556

Half K, Soetendorp B (1998) Small states and the europeanization of public policy. In: Longman (ed) Adapting to European integration: small states and the European Union. Routledge, New York

Hampton MP, Christensen J (2002) Offshore pariahs? Small island economies, tax havens, and the re-configuration of global finance. World Dev 30(9):1657–1673

Harris P (2013) Corporate tax law: structure, policy, and practice. Cambridge University Press, Cambridge

Kaiser W, Starie P (eds) (2005) Transnational European Union: towards a common political space. Routledge, New York

Koh HH (2005) What transnational law matters. Penn State Int Law Rev 24:745–753

Lahey KA (2015) Uncovering women taxation: the gender impact of detaxation, tax expenditure, and joint tax/benefit units. Osgoode Hall Law J 52:427–660

Likosky M (2002) Transnational legal processes: globalisation and power disparities, vol 9. Cambridge University Press, Cambridge

McLean D (1985) Vulnerability: small states in the global society. Commonwealth Secretariat, London

Panayi CHJI (2015) The compatibility of the OECD/G20 base erosion and profit shifting proposals with EU law. Bull Int Tax 70(1/2) (23 pages). Available at SSRN: https://ssrn.com/abstract= 2697511

Robinson WI (2004) A theory of global capitalism: production, class, and state in a transnational world. JHU Press, Baltimore

Sharman J (2005) South Pacific tax havens: from leaders in the race to the bottom to laggards in the race to the top? Account Forum 29(3):311–323

Vlcek W (2008) Offshore finance and small states: sovereignty, size, and money. Palgrave MacMillan, London

Walt SM (1987) The origins of alliances. Cornell University Press, New York

Chapter 6
The Ongoing Legal Status of Low-Lying States in the Climate-Changed Future

Alberto Costi and Nathan Jon Ross

6.1 Introduction

In October 1987, in light of the Brundtland Report published a few months earlier, the then-President of the Maldives, Maumoon Abdul Gayoom, spoke before the United Nations (UN) General Assembly.[1] He warned that a rise in sea levels could lead to the 'death' of his country and other low-lying atoll states, which are extremely vulnerable and uniquely susceptible to the effects of climate change.[2] In his view, their entire territories could be irremediably affected as a result of sea level rise, salinisation of fresh water and storm surges, thus threatening their very existence as sovereign states under international law. Thirty years on and the increasing corpus of climate change data has not sufficed to incentivise the international community to address the complex legal issues surrounding this scenario.[3] In particular, the consequences on the status of states under international law have not yet been resolved.[4]

This chapter examines this statehood question. It considers existing norms of international law and how they may apply in determining the status of a state that no longer has habitable territory. These norms, it must be noted, were largely

---

[1]World Commission on Environment and Development (1987).

[2]Gayoom (1987).

[3]That data includes both the science of rising sea levels and increasing weather extremes, as well as the clear understandings of the human rights and other risks for affected peoples.

[4]Former I-Kiribati President, Anote Tong, explained: 'The question as to what happens to our sovereignty? I don't think anybody has the answer': ABC News (2 April 2007) Kiribati's President: 'Our Lives Are At Stake'. ABC News. Available at http://abcnews.go.com/WNT/story?id=3002001&page=1. Accessed 7 September 2016.

A. Costi (✉) • N.J. Ross
Faculty of Law, Victoria University of Wellington, Wellington, New Zealand
e-mail: alberto.costi@vuw.ac.nz; nathan.ross@vuw.ac.nz

© Springer International Publishing AG 2017
P. Butler, C. Morris (eds.), *Small States in a Legal World*,
The World of Small States 1, DOI 10.1007/978-3-319-39366-7_6

developed in the context of the creation of new states. There is currently no international law directly addressing the continuity or otherwise of a state if its entire territory is rendered uninhabitable due to environmental changes. The central thesis of this chapter, however, is that these low-lying states should, if this scenario eventuates, still retain some form of legal status, albeit ex situ.

To advance this thesis, and following some preliminary issues of terminology (Sect. 6.2), this chapter first summarises the compounding environmental changes confronting low-lying states and describes the scenario this presents for them (Sect. 6.3). Secondly, the adaptation elements of the current international framework on climate change are found to have the potential to assist low-lying states (Sect. 6.4). Thirdly, the prospects of four potential outcomes for the legal personality of these states are examined: full statehood (Sect. 6.5), sovereignty (Sect. 6.6), lesser legal personality (Sect. 6.7), and no legal personality (Sect. 6.8). The worst outcome for the affected states would be to lose all status and rights under international law, but this chapter argues that the other options are available. Finally, a case is made for regional and international dialogue on these legal issues (Sect. 6.9) before offering a few concluding remarks (Sect. 6.10). The importance of state recognition and the complexity of the legal issues are emphasised throughout the analysis, drawing attention on the need for the international community to proactively consider the issues and the legal fate of low-lying states.

6.2 Preliminary Issues

It is commonly said that international law provides a vocabulary for international relations.[5] It is incumbent upon international legal scholars and practitioners, then, to promote language that is constructive and duly sensitive to affected peoples' circumstances. In the context of low-lying states, problematic language has been identified by Rayfuse:[6]

> [F]ar from protecting the rights of persons displaced due to sea-level rise, the use of the essentially negative concepts of refugee and forced migration law serves only to conclusively disempower the persons being displaced.

As well as 'refugee' and 'forced migration', low-lying states themselves are referred to in the literature as 'disappearing states'.[7] The negative connotations of this terminology are problematic per se, but certain problems result from its use.[8] To begin with, there is a dichotomous power dynamic embedded in this language. This is especially apparent when countries that might partner with a 'disappearing

---

[5]Klabbers (2013), p. 315; Shaw (2014), p. 5; Crawford (2012), p. 15.
[6]Rayfuse (2011), p. 284.
[7]McAdam (2010); Rayfuse (2009); Yamamoto and Esteban (2014), p. 2.
[8]UNHCR (2011), at [30].

state' are referred to as 'host states', emphasising their relative position of choice.[9] This power dynamic—apparently constituted in and supported by the international legal discourse—risks exacerbating a sense of disempowerment already present in low-lying states from climate change itself.

Employing troublesome language might also perpetuate a systemic weakness and trigger a sense of reignited colonialism. Through a third world approach to international law (TWAIL), Chimni identifies an 'alienation of international law from the peoples of the third world' and (in a relocation, refugee context) an 'ongoing clash between the humanity of the victims and the rights of sovereign states to exclude [which] reflects the estrangement of international law from its final subjects'.[10] Perpetuating a known problem should be avoided. That said, even ignoring insights from TWAIL, entrenching a power imbalance through dichotomous language undermines a fundamental principle of international law: the sovereign equality of states.[11] The rights of low-lying states and their populations can, as will be explained later, be maintained ex situ. Through formal or informal arrangements, this can be achieved whilst maintaining other states' rights to territorial integrity and political independence.[12]

A final note is on the particular phrase 'disappearing state'. This phrase implies a legal outcome, since statehood is the manifestation of a legal status. Presuming that statehood will 'disappear' is almost certainly erroneous, as will be argued later. This phrase, among all others, carries the greatest risk for the continuity of states and should hence be avoided. Similar language is employed in political discourse by leaders of low-lying states who say that their whole nations are at risk, or that their sovereignty is at risk.[13] This language also implies a legal outcome, but there are no rules or precedents that might suggest this result is even possible, let alone likely. Although such political commentary is well-intended, it may become a self-fulfilling prophecy that could otherwise be prevented. Again, any language that presumes such a negative and legally-unfounded result should be avoided.

In light of this discussion, this chapter refers to low-lying states by this simple descriptive moniker. It is based on an objective fact of their topography that links these states to a common issue without implying any weakness, or any loss of sovereignty or legal personality.[14] In turn, those states that might assist the

---

[9]Hodgkinson et al. (2010); Wilcox (2015); Wyman (2013), p. 187.

[10]Chimni (2007), pp. 500 and 507.

[11]Charter of the United Nations, art 2(1); VCLT at preamble, seventh recital.

[12]Charter of the United Nations, art 2(4).

[13]For example, 'President H.E. Anote Tong addressed the General Assembly of the United Nations...and stressed that it was the Government's desire to "maintain our homeland and our sovereignty"': Office of the President of Kiribati (26 September 2009) President tells of greatest challenge. Available at http://www.climate.gov.ki/2009/09/26/president-tong-addresses-the-un-general-assembly/. Accessed 7 September 2016.

[14]Contrast Rayfuse (2011), p. 284: 'As the territory of a threatened island State disappears beneath the waves, the criteria of territory will no longer be met and the claim to statehood will fail'.

low-lying states are referred to as partner states, implying only—as the *Oxford English Dictionary* describes—that they are 'tak[ing] part in an undertaking with another'.[15]

6.3 The Climate Change Scenario for Low-Lying States

The ability of the peoples of low-lying states to remain and thrive on their territories is being undermined as a result of climate change and its consequences.[16] This challenge is commonly conceived of as 'sinking islands', since the highest point is normally only a few metres above sea level.[17] Due to thermal expansion of oceans and the melting of terrestrial ice and snow, sea level has risen and will continue to rise.[18] Indeed, the rate of sea-level rise is increasing[19] and it is worse in the tropical Pacific, where many of the low-lying states are situated, because the rate of increase in the region is up to four times higher than the global average.[20]

However, the low-lying states are confronted by a much wider range of compounding climate change-related problems and it is worth considering them together for a more holistic understanding of their environmental situations.[21] Low-lying states are confronted by more frequent and severe weather extremes, such as rainfall events and heatwaves;[22] flooding and inundation from sea-level rise, extreme weather events such as cyclones, or both;[23] marine water pollution and salinisation of water supplies, agricultural lands and fresh water ecosystems;[24] and erosion of coastlines and coastal developments.[25] Furthermore, bleaching and

---

[15]Soanes and Stevenson (2009), p. 1044.

[16]A study of peer-reviewed literature on global warming found that 99.99 per cent of published scientists agree that climate change is occurring and is anthropogenic: Powell (2016).

[17]The low lying atoll states at particular risk include Kiribati (consisting of 33 atolls), the Marshall Islands (29 atolls and five islands), Tuvalu (nine low-lying islands), and the Maldives (1190 low-lying atoll islands): see Levine (2009) and Ministry of Environment, Energy and Water (2007).

[18]Stocker et al. (2013), p. 1139.

[19]Stocker et al. (2013), p. 1139; Hansen et al. (2015).

[20]Field et al. (2014a) at 29.3.1.1. The variability in the extent of sea-level rise is due to local geology and changes in ocean currents: Fitzpatrick (2013).

[21]Note that these problems are generalised here and that natural systems are complex and there are some naturally occurring processes that mitigate some of these events. However, the processes that mitigate these impacts are generally outstripped by the climate change effects that create these effects so that the overall effect is clearly negative: see generally Field et al. (2014a), chapter 29: Small Islands.

[22]See generally Stocker et al. (2013); McLeman (2008), p. 11.

[23]Field et al. (2014a), at 29.3.1.1.

[24]Barnett and Adger (2003), p. 325.

[25]Field et al. (2014a), at 29.3.1.1; McLeman (2008), p. 11.

other damage to coral reefs compromise the ecosystem services they provide, for example, protecting island shores and providing habitat for marine species that are important to subsistence.[26] Similarly, the greater depth of the sea damages mangroves and sea grasses, thereby compromising their supply of ecosystem services, such as providing foods, materials, and habitat for other species that are important for subsistence.[27]

These environmental changes have human health impacts: direct mortality and injury from extreme weather events;[28] increased incidences of diseases such as malaria and dengue fever;[29] diseases from exposed landfill and burial sites following floods and inundation events;[30] and compromised health from lack of access to freshwater and adequate nutrition.[31]

Whilst there are domestic issues that also incentivise relocation, such as high unemployment, the scale of these compounding environmental changes is clearly enormous and the developing low-lying states simply lack the capacity to protect themselves. Ideally, the need for relocation would be prevented, but efforts to mitigate greenhouse gases continue to be inadequate.[32] Furthermore, although there are options to adapt to climate change, these are likely to be unrealistic and unaffordable in the long-term.[33] The Maldives has looked into island protection, but costs of US$6 billion dollars for coastal protection, or of US$500 million to US$1 billion for elevating islands by one metre, were deemed too expensive.[34] Low-lying states in the Pacific, relying on an annual gross domestic product (GDP) ranging from US$27 million in Tuvalu to US$644 million in Vanuatu,[35] would be similarly unable to afford such measures.

In other words, while acknowledging peoples' desires to remain where they are,[36] in situ adaptation is unlikely to provide a comprehensive long-term solution to the scale of adverse effects summarised above. The Intergovernmental Panel on Climate Change (IPCC) has noted that 'the very existence of some atoll nations is threatened by rising sea levels'[37] and that 'land inundation due to sea-level rise poses risks to the territorial integrity of small-island states'.[38] Thus, this is the

---

[26]Ibid at 29.3.1.2.

[27]Ibid.

[28]Ibid.

[29]Ibid.

[30]Foster (2014).

[31]Field et al. (2014a), at 29.3.3.2.

[32]King and Henley (2016).

[33]The potential for 'astronomical' adaptation costs were a key reason why the United States would not accept references to liability and compensation in the *Paris Agreement*: Little (2015).

[34]Shaig (2006), p. 15.

[35]United Nations UN Data (2016a).

[36]Valemie (2016).

[37]Field et al. (2014a), at 29.1.

[38]Field et al. (2014b), at 21.

climate change scenario for low-lying states: since in situ adaptation cannot match the scale of environmental change, ex situ solutions will become necessary. It is against this scenario that the law is analysed in this chapter.

6.4 Current International Framework on Climate Change

Before considering the legal personality of low-lying states in this scenario, it is useful to examine the current climate change regime under international law. Importantly, it does not (yet) consider the continuity of states, but this may change. International climate change law is framed around a cascade of failure, starting with the assumption that the world will mitigate the problem. When that fails, we will adapt to climate change. When that fails, we will look towards funding for loss and damage. At the next phase, looking beyond in situ resolutions, there is work on developing a 'climate change displacement coordination facility', which was referred to in a draft of the Paris Agreement.[39] Should this trend continue, international climate change discussions may eventually consider the issue of retention of statehood, since the physical undermining of whole states represents an extreme failure of the international community and constitutes an entirely novel problem for public international law.

Although state continuity is not directly referred to in any of the binding climate change treaties, it is not impeded by any of their provisions either. In fact, continuity is supported by two helpful elements. The first is adaptation itself. It has been suggested that the implementation of adaptation provisions has been hampered by a lack of agreement about the meaning and scope of adaptation.[40] However, any definition could be improperly limiting, given the vast array of activities required for successfully adapting diverse human and natural systems to the multitude of distortions and transformations wrought by climate change.

The only clear delineation of the concept of adaptation is that it must be relative to 'the adverse effects of climate change', which are defined in the United Nations Framework Convention on Climate Change (UNFCCC) as:[41]

> [C]hanges in the physical environment or biota resulting from climate change which have significant deleterious effects on the composition, resilience or productivity of natural and managed ecosystems or on the operation of socio-economic systems or on human health and welfare.

Given the strong presumption of continuity of states in international law[42] and the primacy of sovereignty of states,[43] along with the multitude of risks for human

---

[39]UNFCCC (2015a), pp. 31–32. See also Wentz and Burger (2015).

[40]Yamin and Depledge (2004), p. 213.

[41]UNFCCC (1992), article 1, 'adverse effects of climate change'.

[42]Crawford (2006), p. 34.

[43]Charter of the United Nations, art 2(1) and UNGA Declaration on Friendly Relations (1970), p. 124.

beings who are at the centre of international law, it is undeniable that the potential disappearance of a state is a significantly deleterious effect.

In the *Paris Agreement*, the scope of Article 7 regarding adaptation is more comprehensive and aspirational than the coverage that adaptation was given in previous decisions of the Conference of the Parties.[44] Those decisions often referred to adaptation as 'including' 'financing', 'technical assistance' and 'capacity building'. In Article 7, however, '[p]arties recognize that the current need for adaptation is significant and that … greater adaptation needs can involve greater adaptation costs'.[45] In particular, Article 7 provides that:[46]

> Parties recognize the importance of *support* for and *international cooperation* on adaptation efforts and the importance of taking into account *the needs of developing country Parties*, especially those that are particularly vulnerable to the adverse effects of climate change.

Only states are Parties to the *Paris Agreement*. Therefore, 'the needs' of states—'especially those that are particularly vulnerable to the adverse effects of climate change'—must be taken into account as concerns the 'support' and 'cooperation' of the international community under this Agreement.[47] The state's principal need is to continue in its existence so that it can protect its citizens, domestically, regionally and internationally. Support and cooperation, therefore, must be to that end: enabling the continuity of the state.

The second helpful element from the international climate change sphere is the emergence of work on 'planned relocation'[48] and the proposal for a 'climate change displacement coordination facility'.[49] At the sixteenth Conference of the Parties, the Cancun Adaptation Framework was adopted in which all Parties were:[50]

> Invite[d] … to … undertak[e] … [m]easures to enhance understanding, coordination and cooperation with regard to climate change induced displacement, migration and planned relocation, where appropriate, at the national, regional and international levels.

This invitation indicates that an ad hoc approach (that would almost certainly be the default, reactive, maladaptation approach) is undesirable and, therefore, an erudite process is anticipated instead. As for the displacement coordination facility, reference to it was removed from the text of the final *Paris Agreement*.[51] However, this does not prevent Parties from reaching an agreement on such a system at a later time,[52] which might build on closely-related work, for instance, the recent

---

[44]UNFCCC (2015b) Paris Agreement, art 7.

[45]Ibid., art 7(4).

[46]Ibid., art 7(6) (emphasis added).

[47]Ibid., at preamble, fifth recital.

[48]UNFCCC (2010) Cancun Adaptation Framework, at [14(f)].

[49]UNFCCC (2015a).

[50]UNFCCC (2010) Cancun Adaptation Framework, at [14(f)].

[51]Milman (2015).

[52]See Randall (2015).

Guidance on Planned Relocation from the United Nations High Commissioner for Refugees (UNHCR).[53] In the interim, states can innovate and develop their own arrangements, bilaterally, sub-regionally, or regionally, whilst waiting for the international community to catch up.

On this work stream, it is interesting to note the phrase 'induced displacement, migration and planned relocation'. Displacement and migration have particular legal meanings and connotations. In contrast, planned relocation seems to be a new concept, thereby untrammelled from existing legal norms and providing a degree of flexibility that might, inter alia, incorporate measures for entire states.

6.5   Legal Status of a Low-Lying State: Statehood

The novel factual situation that appears to be emerging for low-lying states has implications for many areas of public international law. This chapter focuses on the fundamental issue of the states' legal status. Will they maintain statehood under international law? If not, can some form of sovereignty be retained? If that is not possible, can the low-lying states possess an alternative legal personality? Statehood, sovereignty and legal personality are different and separate attributes which will be discussed subsequently. It is argued here that these attributes, although interlinked, may be held separately. Before examining these legal statuses, the importance of statehood will be summarised to highlight what is at stake in decisions that affect the low-lying states.

6.5.1   The Importance of Statehood

The Peace of Westphalia in 1648 is often referred to as the beginning of the emergence of the sovereign state in the international system. From then on, the Westphalian state (now simply referred to as a state) has been based on, inter alia, the principle of territory, the integrity of which is threatened by the climate change scenario presented in Sect. 6.3 above.[54] To date, the nation-state remains the key actor in international law,[55] which underscores the importance of statehood and its continuity.

Statehood offers certain privileges and rights while also imposing duties and obligations. Traditional legal theory holds that the state is the main actor in the international system and that states are equal and sovereign.[56] Further, only states

---

[53]UNHCR (2015).

[54]Krasner (2001), p. 17.

[55]Besson (2009), p. 360.

[56]Charter of the United Nations, art 2(1).

are *full* subjects with *full* capacity under international law. For example, only states can bring cases before the International Court of Justice (ICJ).[57] States also have other exclusive rights under international law, including the continued effect of existing treaties and the capacity to enter into new agreements, the right to exploit natural resources, and the right to exclusive power within its jurisdiction. Under article 4 of the Montevideo Convention on the Rights and Duties of the State (the Montevideo Convention), the privileges of statehood include being:[58]

> [J]uridically equal, enjoy[ing] the same rights, and hav[ing] equal capacity in their exercise. The rights of each one do not depend upon the power which it possesses to assure its exercise, but upon the simple fact of its existence as a person under international law.

The Montevideo Convention then adds to this by stating that 'the fundamental rights of states are not susceptible of being affected in any manner whatsoever'.[59] States possess these rights simply by having met the statehood criteria. The reasoning behind the equal sovereignty and full legal capacity of states is that states make international law and other international subjects derive their legal personality from states.[60] State recognition is important, and states can, therefore, create new legal principles in regard to the scenario emerging for low-lying states.

Another important function of the state is the diplomatic protection of, and consular assistance for, citizens who are injured abroad.[61] It may be especially important for low-lying states that have relatively large numbers of citizens relocating to other states. This is currently occurring in the Pacific, with large numbers of people emigrating from affected Pacific Islands. Many go to New Zealand, where the annual Pacific quota allows entry to skilled migrants from some of the Pacific Island nations.[62] However, there will likely be a proportion of the population remaining on the original islands that will eventually need diplomatic protection to prevent human rights abuses abroad following final relocation. New Zealand, for example, acknowledges that final relocation will require humanitarian response efforts to ensure human rights and human security are upheld.[63] The diplomatic protection and consular assistance that low-lying states can currently offer will become unavailable if they are no longer recognised as states.

The purpose of that summation is simply to emphasise that, with respect to the statehood question, the first priority for low-lying states is the maintenance of their legal personality. The likelihood of this is analysed here relative to orthodox criteria for statehood and to state recognition.

---

[57]Statute of the International Court of Justice, art 34.

[58]Montevideo Convention, art 4.

[59]Ibid., art 5.

[60]Harrison (2007), p. 161: '[I]f law makes states, it is states that make the law that makes states . . . those already in the family allow new members to enter'.

[61]Shaw (2008), p. 823; International Law Commission (2006).

[62]Immigration New Zealand (2016), Available at www.dol.govt.nz. Accessed 24 September 2016.

[63]New Zealand Government (2009), p. 3.

6.5.2 Requirements of Statehood

The requirements for establishing statehood have been codified in the 1933 Montevideo Convention as follows: (1) a defined territory, (2) a permanent population, (3) a government, and (4) the capacity to enter into relations with other states.[64] The Montevideo Convention is recognised as codifying the customary international law relevant to our understanding of what statehood entails.[65] Furthermore, a vast jurisprudence and discussion have since emerged on the more precise elements of the four criteria required for statehood as listed in the Montevideo Convention. These criteria are examined in turn below.

It should be noted that these criteria are for *establishing* statehood. Continuity of already-established statehood is a separate legal question, which is examined later. The purpose of reviewing the Montevideo Convention criteria, then, is simply to identify the key factual risks as they relate to orthodox international law on statehood. This is especially important because the law on continuity of statehood is not codified and, therefore, less clear.

6.5.2.1 Defined Territory

Territory under international law consists of land territory, internal waters, territorial sea and air space above this territory.[66] If all the islands of a low-lying state become uninhabitable, will the statehood requirement of territory still be met, such as through the other remaining components of its erstwhile territory (such as internal waters, territorial sea and air space above this territory)? An 'island' is defined by the UN Convention on the Law of the Sea (UNCLOS) as 'a naturally formed area of land, surrounded by water, which is above water at high tide'.[67] In a 2001 judgement, the ICJ used that definition to help decide if the disputed islands constituted territory and, therefore, were able to be used in the drawing of equidistance lines. It held that:[68]

> In accordance with Article 121, paragraph 2, of the 1982 Convention on the Law of the Sea, which reflects customary international law, islands, regardless of their size, in this respect enjoy the same status, and therefore generate the same maritime rights, as other land territory.

Predictably, then, islands, including coral atolls, are part of a state's land territory. However, according to the UNCLOS definition, if an atoll is inundated

---

[64]Montevideo Convention, art 1.

[65]Shaw (2014), p. 198.

[66]Duursma (1996), p. 116.

[67]United Nations Convention on the Law of the Sea, art 121(1). For a recent exposition on what comprises an 'island', see *In the Matter of the South China Sea Arbitration (Philippines v China) (Award)* [2016] PCA Case No 2013–19 at [473]–[553].

[68]*Maritime Delimitation and Territorial Questions between Qatar and Bahrain* [2001], at [185].

and disappears, or is flooded at high tide, it will no longer constitute an 'island'. As such, it is unlikely to be considered land territory, as required by UNCLOS. If this is to occur to all islands of a low-lying state, there will no longer be any land territory. However, UNCLOS uses its definitions of territory and island for the purposes of allocating maritime resources and defining maritime boundaries, not for statehood.

An arbitration before the Permanent Court of Arbitration may also help to understand the meaning of territory in the context of islands. In the *Island of Palmas* case, the sole Arbitrator, Max Huber (at the time the President of the Permanent Court of International Justice) described the territory of Palmas as 'a somewhat isolated island, and therefore a territory clearly delimited and individualised'.[69] Earlier in the award, Arbiter Huber stated that 'sovereignty in relation to a portion of the surface of the globe is the legal condition necessary for the inclusion of such portion in the territory of any particular State'.[70] The island of Palmas was also inhabited by a population, which he considered a relevant factor: it meant that acts of administration occurred, thus suggesting sovereign control.[71] Three key elements of territory are discernible from these *Island of Palmas* quotes: delimitation, sovereignty, and population.

The importance of delimitation emanating from the *Island of Palmas* case has been reiterated by legal writers and governments. Delimited borders are not essential in international law. Arguably, however, clear and separate geographic territory allows for the clear identification of a state and the physical area it controls.[72] Even governments-in-exile depend on a physical and definable territory upon which to base their legitimacy.[73]

The territory of a low-lying state could remain above water at high tide but be uninhabitable because of other environmental changes. If it is flooded at high tide, or entirely inundated, it will no longer have clear physical boundaries and, consequently, delimiting territory could create difficulties. For most low-lying states, however, maritime boundaries are secured by the deposit of coordinates and nautical charts with the UN,[74] often on the basis of bilateral treaties[75] and domestic

---

[69]*The Island of Palmas case* (1928), p. 855.

[70]Ibid., p. 838.

[71]Ibid., p. 855.

[72]Taylor (1997), p. 758.

[73]McAdam (2010), pp. 7 and 11.

[74]'Deposit by Kiribati of list of geographical coordinates of points, pursuant to article 16, paragraph 2, article 47, paragraph 9, and article 75, paragraph 2, of the Convention' (2 January 2015) M.Z.N.111.2015.LOS; 'Deposit by the Republic of the Marshall Islands of geographical coordinates of points, pursuant to article 16, paragraph 2, article 47, paragraph 9, and article 75, paragraph 2, and article 84, paragraph 2 of the Convention' (3 May 2016) M.Z.N.120.2016.LOS; and 'Deposit by Tuvalu of lists of coordinates of points, pursuant to article 16, paragraph 2, article 47, paragraph 9, and article 75, paragraph 2, and article 84, paragraph 2, of the Convention' (4 September 2013) M.Z.N.98.2013.LOS.

[75]For example, Agreement between Tuvalu and Kiribati regarding their Maritime Boundary (29 August 2012).

legislation.[76] The ongoing validity of these boundaries is normally a result of the rules applicable to succession, or the terms of relevant treaties (that is, delimitation treaties). In the climate-changed future, the persistence and legal effect of those UN deposits, and the treaties and legislation on which they are based, will be a consequence of decisions on statehood continuity. In other words, just as the delimitation treaties may or may not help to maintain statehood, the (dis)continuity of statehood will determine the future of those boundary deposits. Delimitation boundaries may be a matter directly addressed in a bilateral agreement or later in legislation. In other words, the delimitation aspect of the territory criterion of statehood is not conclusive and could, in fact, be maintained.

The second point from the *Island of Palmas* decision pertains to sovereignty as one of three key elements for determining whether the territory criterion of the Montevideo Convention continues to be satisfied. Sovereignty is not a criterion for establishing statehood, but rather it is a state's right that arises as a consequence of that legal personality being established.[77] The *Island of Palmas* case (insofar as the sovereignty element of the territory criterion is concerned) determined that the right of sovereignty over territory could only be upheld if it was effectively exercised (*animus occupandi*).

Scholarly debate has taken place as to the interpretation of sovereignty as a requirement to satisfy the territorial criterion for statehood. Whether sovereignty as envisaged by Arbiter Huber persists with respect to the novel climate change scenario for low-lying states is, of course, untested. Crawford has argued that the criterion for effective government is applied more strictly to situations of establishing states than to continuity of states.[78] One might argue that this looser application to existing states might translate also to other statehood criteria and indicia, including undisputed territories of a sovereign nation that become uninhabitable due to anthropogenic climate change. Again, then, this aspect of the territory criterion of statehood is not conclusive.

The third element of Arbiter Huber's conception of territory is a population and this is dealt with separately below. On the territory criterion overall, some commentators have assumed that loss of (habitable) territory, in itself, results in the discontinuation of the sovereign state entity.[79] For example, the former President of the Federated States of Micronesia, Leo Falcam, said 'sea level rise ... [is a] grave security threat to our very existence as ... nation-states'.[80] Rayfuse agrees that, as territory disappears, the criterion of territory will not be met and she presumes that this automatically extinguishes the legal personhood of the state.[81] Ryngaert and Sobrie believe that central to the Montevideo Convention is the principle of

---

[76]For example, Republic of the Marshall Islands Maritime Zones Declaration Act 2016.

[77]Montevideo Convention, art 4.

[78]Crawford (2006), at 59.

[79]UNHCR (2009), pp. 1–2; UNGA (2008).

[80]Dupont and Pearman (2006), p. 47.

[81]Rayfuse (2011), pp. 282 and 284.

effectiveness, and that a low-lying state cannot be effective without physical territory, due to forced displacement of the population and inability of the government to control a particular territory or carry out the functions of the state.[82]

It is not a foregone conclusion that a low-lying state without habitable territory has no chance of meeting the delimitation or sovereignty elements of the Montevideo Convention criterion of territory. There is simply no authority for a conclusion with such drastic consequences. Furthermore, both delimitation and sovereignty can be addressed in ways that maintain a legal tie to territory. As indicated, delimitation over uninhabitable land territory and over territorial sea can, for example, be enabled through existing and new treaties.

Similarly, sovereignty, as the right and power of independence, can take many forms, including forms that are to-date unforeseen. As the Permanent Court of International Justice (PCIJ) stated in *Legal Status of Eastern Greenland*, sovereignty requires 'the intention to act as sovereign, and some actual exercise or display of such authority'.[83] The means and efficacy of climate change adaptation measures could very well define new forms of exercising sovereignty, whilst almost certainly drawing on previous international experiences. Sovereignty may be exercised over the current delimited zones through a legal arrangement with a partner state, and it may be exercised—in some form or another—over either territory or property currently within the territory of another state. To assume that sovereignty is lost with existing territory retards the potential for creative solutions to an entirely novel problem. Indeed, nothing in law prevents ex situ continuity of sovereignty.

The potential for innovative means of maintaining a state's links to its territory has been seen, for instance, in situations where the state loses effective control due to breaches of peace and security. For example, following occupation by Nazi Germany and Russia, the Government of Poland operated in exile in France and then in the United Kingdom during World War II.[84] Losing control of territory in an illegal situation of aggression does not translate into loss of statehood. Indeed, the opposite result—that statehood persists—is presumed. This is a measure intended to ensure the failure of such illegal activity, thereby acting as a disincentive for such actions. The state's agents tend to operate from outside the territory of their state because of the tumultuous circumstances.

There are two key differences between situations such as that of a government-in-exile, for instance World War II Poland, and a low-lying state. The first is that situations of breaches of international peace and security tend to be temporary, as opposed to the permanent loss of habitable territory due to climate change. The second key difference is the cause of the situation, one being aggression (clearly an

---

[82]Ryngaert and Sobrie (2011), p. 472.

[83]*Legal Status of Eastern Greenland* (1933), p. 46.

[84]Flemming (2014), p. 78. The relocation to the United Kingdom was supported by a bilateral accord, *Respecting the Polish Forces in the United Kingdom* AIR 2/4213 and WO 33/2389, National Archives, London. See Peszke (2011), pp. 648–650.

illegal act), while the other is environmental degradation. Despite these differences, there is a presumption of continuity of the state. As Crawford explains, 'there is a substantial body of practice protecting the legal personality of the state against extinction, despite prolonged lack of effectiveness'.[85]

Why is this so? On its own, avoiding the threat or use of force is a minimal, negative standard, and peace is more than that. It has positive aspects, such as cooperation, social justice, wellbeing and health, and access to democratised social institutions.[86] Thus, what ultimately sits underneath the defence of statehood is not merely the prevention of the use of force, but instead the protection of individual human rights and the collective right to self-determination against illegal situations. An Expert Group on Global Climate Obligations has convincingly argued that the failure of the international community 'to achieve ... stabilization of greenhouse gas concentrations in the atmosphere at a level that would prevent dangerous anthropogenic interference with the climate system'[87] is an illegal situation.[88] If so, then the loss of statehood in this climate change scenario may constitute a departure from practice whereby statehood is otherwise presumed to persist when challenged by a wrong at international law.

In any event, to presume that loss of territory suffices to extinguish the legal personality of a state is an exceptionally bold and dangerous idea. History has shown time-and-again the devastating effects for peoples (and for the broader international community) when deprived of their national identity, culture, custom and language. Whilst the scenario considered in this chapter is based on the facts of loss of existing habitable territory, there is a range of possible futures for low-lying states ex situ that address delimitation and sovereignty.

6.5.2.2 Permanent Population

The issue with this Montevideo Convention statehood criterion is whether the requirement for a permanent population will be undermined by the climate change scenario. Logic suggests that permanent population will be the first criterion at risk since people will leave before the territory becomes completely uninhabitable.[89] For example, Kiribati's former President, Anote Tong, announced plans in 2010 to organise the migration of its population of about 110,000 to other states as the situation worsens.[90] In 2008, Tuvalu's then-Prime Minister, Apisai Ielemia, issued

---

[85]Crawford (2006), p. 132.

[86]Anderson (2004).

[87]UNFCCC (1992), art 2.

[88]Expert Group on Global Climate Obligations (2015).

[89]McAdam (2010), p. 1.

[90]Office of the President of Kiribati (2010).

an official request to the Australian Government to give Tuvalu a parcel of territory for the purpose of relocating and re-establishing Tuvalu.[91] This was refused.[92]

'Population' in the statehood context comprises the individuals who inhabit the territory permanently.[93] The size of the population is irrelevant.[94] However, the permanence of the population is germane, as is 'the maintenance of an essentially permanent form of community life in the sense of sharing a common identity'.[95]

Brownlie believes that the population criterion is intended to be interlinked with the requirement for territory, connoting a stable community in control of a specific area.[96] Hence, it appears that, *if* the territory becomes *entirely* uninhabitable and *all* people relocate to the territories of other states, the population criterion will not be met. The implication of the italicised words is that this criterion will continue to be satisfied as long as any small number of people from a low-lying state continue to live there. However, if all people leave—the basis of this climate change scenario—then, prima facie, this criterion will no longer be satisfied.

However, if the affected peoples (or a majority of that population) relocate to a new land (whether as territory ceded from another or, more likely, as property held in private law), then there remains a population with a form of community life: a nation—a relatively homogenous peoples—linked together by ethnicity, national identity, genealogy, culture, traditions, history and language. This demonstrates the risks associated with fragmenting the populations of low-lying states. If their relocation relies on migration law alone, some people may end up, say, in New Zealand, others in Fiji, and others in Australia. There is a clear risk, in such situations, of gradual integration of the affected peoples into the majority community, whether quite quickly or over a few generations. This is why unplanned relocation amounts, in effect, to a policy of assimilation and the denial of the collective right to self-determination. It also demonstrates that it is possible to have a population—a peoples—with a form of community life that remains as a permanent population, albeit in a new territory.

It is perhaps worth noting that, although a territory may become uninhabitable, there is nothing in international law that requires it to be habitable, except for the purpose of maritime delimitation. It is theoretically possible, then, for a state to persist in a form of environmental exile. Thus, the territory made uninhabitable by the adverse effects of climate change remains the territory of the state *in abstentia*. The inclusion of uninhabited land within a state's territorial claims certainly has precedent. New Zealand, for example, has numerous islands that form part of its territory, some of which have never been inhabited, such as the Solander Islands (although five sealers were inadvertently marooned there for five years).[97]

---

[91]Displacement Solutions (2010), p. 20.

[92]Rayfuse (2011), p. 282.

[93]Duursma (1996), p. 117.

[94]*In re Duchy of Sealand* [1978], p. 687.

[95]Ibid.

[96]Brownlie (2008), pp. 70–71.

[97]Wikipedia. Solander Islands. Available at https://en.wikipedia.org/wiki/Solander_Islands. Accessed 24 September 2016.

In summary, in the scenario of total loss of habitable territory, the permanent population criterion will not be met in situ. The issue, however, must be measured against the legal requirements for continuity of statehood. In other words, can the low-lying states persist ex situ, where the entirety or a majority of the population relocates? The matter of options for facilitating ex situ continuity is beyond the scope of this chapter, but it is important that the issue is not disposed of without at least indicating that there is potential and that a proper analysis of options is desirable.[98]

6.5.2.3 Government

The third criterion for statehood is the requirement for a government, which reflects the need for a state to have international representation, among other things. Effective government also refers to the ability of the government to control its territory and people, its right to exercise such authority, and the exclusivity of that right.[99] Does this criterion require simply a government in power, or an effective government? Crawford argues for the latter.[100]

A concern, then, is the ability of the governments of low-lying states to continue to fulfil their basic functions. Park is concerned that the low-lying states' dependency on funding from other states may limit their ability to guarantee basic rights and services for their citizens.[101] Similarly, Sinclair notes that the effects of climate change risk leading to internal instability and state failure.[102] A failed state is one that is unable to ensure domestic order by monopolising the use of force in its territory.[103] Kahl argues this may occur as a result of the impacts of climate change on government revenue and legitimacy, and on social cohesion.[104] Also, once the territory is no longer habitable, it can be argued that a government cannot be effective if it cannot carry out functions within its territory or rule over the displaced population, especially if it is fragmented.

However, many states today retain statehood without functioning governments. The 2012 Failed States Index listed 10 states unable to provide public services required of government.[105] McAdam also argues this point, citing the situation of the Congo in 1960 as an example.[106] Yet, these 'failed' states still possess statehood. Taking this reality into account, it is likely that what is required is simply a

---

[98]See Burkett (2011).

[99]Crawford (2006), pp. 57–59.

[100]Ibid., p. 33.

[101]Park (2011), pp. 12–14.

[102]Sinclair (2008), p. 41.

[103]Brooks (2005), p. 1160.

[104]Kahl (2006), p. 44.

[105]For instance, Chad, Somalia and Zimbabwe: see Foreign Policy (2016).

[106]McAdam (2010), pp. 6 and 8.

government rather than an effective government. Shaw notes, in respect of Croatia and Bosnia-Herzegovina, that these independent states were recognised by the European Community in spite of the fact that non-governmental forces controlled much of their territories.[107] Supporting these case studies, the Montevideo Convention provides that, '[t]he rights of each [state] do not depend upon the power which it possesses to assure its exercise'.[108] In sum, statehood has continued elsewhere when governments of existing states had limited effectiveness.

This supports the rule that the form of government is irrelevant. The ICJ commented in the *Western Sahara* advisory opinion that:[109]

No rule of international law, in the view of the Court, requires the structure of a State to follow any particular pattern, as is evident from the diversity of the forms of State found in the world today.

Add to this Crawford's conclusion that:[110]

[I]nternational law lays down no specific requirements as to the nature and extent of this [governmental] control [of territory], except that it include some degree of maintenance of law and order and the establishment of basic institutions.

He remarks further that:[111]

[T]here is a distinction between the creation of a new State on the one hand and the subsistence or extinction of an established State on the other. In the former situation, the criterion for effective government may be applied more strictly.

Using these understandings for analysing the third requirement of the Montevideo Convention, one could argue that a low-lying state could very well continue to have a government. For the purposes of the continuity of the existing low-lying states, any form of government—including ex situ government—that is exercising any form or degree of control of its territory—including uninhabitable islands and territorial sea—should suffice in satisfying the government criterion of the Montevideo Convention. The concerns about revenue, securing governmental services and human rights, social cohesion and so on are certainly real and pragmatic issues that would still need to be resolved. For the purpose of determining continuity of state, however, it is far from clear when, or if ever, this criterion would cease to be satisfied. Crawford has argued that 'there is a good case for regarding government as the most important single criterion of statehood, since all the others depend upon it'.[112] If so, then the continuity of statehood may well persist and follow the low-lying states' governments and peoples, rather than be extinguished by the loss of existing territory and permanent in situ population.

---

[107]Shaw (2014), p. 146.

[108]Montevideo Convention, art 4.

[109]*Western Sahara* [1975], at [94].

[110]Crawford (2006), p. 59.

[111]Ibid.

[112]Crawford (1979), p. 56.

6.5.2.4 Capacity to Enter into Relations

The final criterion for statehood according to the Montevideo Convention is the capacity to enter into international relations. Whilst this criterion relies on the willingness of other states to reciprocate, it is also a matter of competence, that is, the state should have the ability to conduct international relations with other states.[113] For Crawford, capacity 'is not a criterion, but rather a consequence, of statehood, and one which is not constant but depends on the status and situation of particular States'.[114] The low-lying states have already shown capacity and acted upon it regularly, most notably in their bustling participation in international climate change negotiations through the Alliance of Small Island States (AOSIS) and the 'high ambition coalition'.[115]

In terms of ongoing capacity and international agreements between low-lying states and other states entered into before relocation, those agreements are presumed to continue.[116] Capacity to enter new agreements, however, will depend on the willingness of other states to enter into relations. Although not a formal requirement, continued state recognition will have practical effects, especially as regards capacity.[117] This challenge may, to some extent, be ameliorated by continued recognition by friendly states, such as other members of AOSIS. The issue of recognition is discussed further below.

6.5.3 Montevideo: New States Only

External strategic intervention and support can help to maintain the criteria for statehood contained in the Montevideo Convention. However, without that intervention, the criteria are unlikely to be maintained by low-lying states with uninhabitable territories and exiled peoples. Regardless, while some academics claim that a state will cease to exist when it cannot meet a criterion,[118] it must be noted that both customary international law and the Montevideo Convention only offer requirements for a *new* state to be established and gain statehood. Neither specifies the requirements for the *continued* existence of states. To put it differently, neither custom nor the Montevideo Convention specify what milestones must be

---

[113]*U.S. Restatement of the Law, Third, Foreign Relations Law of the United States*, Vol. 1, Sec. 201, cited in Grant and Barker (2009), p. 574.

[114]Crawford (1979), p. 47.

[115]Alliance of Small Island States, www.aosis.org and Mathiesen and Harvey (2015). For a summary of the low-lying states' participation in international climate change discourse, see Yamamoto and Esteban (2014), pp. 105–119.

[116]VCLT, arts 54–60. Note certain exceptions including supervening impossibility of performance (art 61) and fundamental change of circumstances (art 62).

[117]Ryngaert and Sobrie (2011), p. 470.

[118]Rayfuse (2011), p. 281.

reached for the status of statehood to be extinguished. In fact, in contrast to this, customary international law includes a principle of continuity of states.[119] Article 6 of the Montevideo Convention codifies this principle: 'recognition [of a state] is unconditional *and irrevocable*'.[120] Traditionally, the principle of continuity of states has meant that, once a state is recognised, recognition cannot be withdrawn. (This differs from recognition of government, which is political and can be altered.) The principle of continuity is based on a strong assumption against the extinction of states once they have been firmly established, which is the case for all low-lying states.[121]

Since the creation of the UN, virtually no involuntary extinction of states has occurred except by purposeful dissolution.[122] Succession and the merging of states can lead to the end of recognition of a state for practical reasons. However, there is nothing to suggest that the principle of continuity will become irrelevant if a low-lying state no longer satisfies a statehood criterion, at least not in an orthodox form. Shearer, for example, makes the argument that once a state is already established, the requirement of territory is not necessary.[123] Grant makes a similar point, that 'once an entity has established itself in international society as a state, it does not lose statehood by losing its territory or effective control over that territory'.[124] These scholars were writing in the context of events such as World War II and the civil war in Somalia, where territory did not disappear but was taken over by another entity, whether temporarily or more permanently. However, it demonstrates the fluidity of the Montevideo criteria for continuity of statehood.

While this chapter considers current norms of international law, these norms do not contemplate circumstances such as those applicable to low-lying states. At the heart of this emerging situation, though, is the pivotal place of states in international law, plus the principle of continuity. Together, these tenets mean that it is erroneous to presume that the legal personality of the low-lying states will be extinguished, especially in the total absence of any rules governing such a critical outcome and process.

6.5.4 Importance of State Recognition

Many principles and norms in existing international law apply to the emerging situations of the low-lying states: statehood, self-determination, sovereign equality of states, and so on. However, since the factual scenario is entirely novel, the

---

[119]*Tinoco Claims Arbitration* (1923), pp. 377–379.

[120]Montevideo Convention, above n 58, art 6 (emphasis added).

[121]Crawford (2006), p. 715.

[122]McAdam (2010), pp. 5–6.

[123]Shearer (1994), p. 85.

[124]Grant (1999), p. 435.

reaction of the international community will be important. International law-making is, after all, vested in states themselves.

Thus, politics will play an important role. Ryngaert and Sobrie highlight that, although recognition of states traditionally relies on a set of basic legal rules, recent practice has illustrated a different reality whereby politics rather than legal norms play the leading role.[125] This impacts upon the rule of law, which involves, inter alia, 'legal certainty, avoidance of arbitrariness and procedural and legal transparency'.[126] For the peoples of low-lying states, their future is less certain if the involvement of unhelpful politics interferes with legal norms, such as the principle of state continuity, that could otherwise be relied upon.

The disappearance of Bermeja Island off Mexico offers an interesting case study of the politics of territory. In 2009, it was reported that Bermeja Island, part of the territory of Mexico, could not be found.[127] The importance of this island is that it extended Mexico's exclusive economic zone. When Bermeja 'disappeared', the United States made strong statements to the effect that, without the island, there is no territory and no maritime claim to that part of the Gulf of Mexico.[128] At stake were valuable oil deposits that were within the exclusive economic zone that extended from Bermeja.[129] That area of the Gulf of Mexico now belongs to the United States. The National Autonomous University of Mexico investigated the island and concluded that it must not have ever existed.[130] Regardless, the early reaction of the United States to the 'disappearance' shows the political importance behind island recognition when exploitation rights are concerned.

Many of the low-lying states have valuable maritime resources and rights based on their territory. If the land territory upon which rights to these resources rely disappears, the territorial sea and exclusive economic zones of low-lying states could be claimed by another state or become part of the high seas. Such occurrence will likely affect whether the territory, and subsequently the low-lying state, is recognised by neighbouring states. This situation will, however, be influenced strongly by the existence of the coordinates and nautical charts of maritime boundaries deposited with the UN and the legal instruments upon which they are based, as discussed above.

Intertwined with this political issue is the question of who enjoys the competency to decide if a low-lying state meets the criteria for statehood. This answer is made clear by chapter II of the UN Charter in the context of gaining UN membership: the General Assembly decides membership of 'peace-loving states' on the advice of the Security Council.[131] However, an entity can be a state without UN

[125]Ryngaert and Sobrie (2011), p. 467.

[126]Annan (2004), at [6]; see UNGA (2012).

[127]Paskal (2010).

[128]Ibid.

[129]Cuen (2009).

[130]Banderas News (2009).

[131]Charter of the United Nations, art 4.

membership. Outside of the UN, the question is more complex. Whilst the Montevideo Convention provides that '[t]he political existence of the state is independent of recognition by the other states', the reality is more complicated.[132] Individual states often believe they have a right to decide whether or not another entity is entitled to statehood. This is illustrated by the entities which are recognised by some but not other states, such as Taiwan (Republic of China), Palestine and the (Republic of) South Ossetia.

The universally-accepted and uncontroversial statehood status of low-lying states today may or may not reduce controversies surrounding the removal of recognition as the climate change scenario unfolds. The low-lying states may, however, argue that withdrawing recognition from an existing state is entirely different from denying recognition of an emerging entity. Even with the latter scenario, Crawford describes the issue as:[133]

> [W]hether the denial of recognition to an entity otherwise qualifying as a State entitles the non-recognizing State to act as if it was not a State—to ignore its nationality, to intervene in its affairs, generally to deny the exercise of State rights under international law.

If this holds true for emerging states, it certainly also does for pre-existing states. Assuming that low-lying states engage in efforts to maintain their statehood ex situ, removal of recognition would be a breach of 'the duty not to intervene in matters within the domestic jurisdiction of any State', contained in the *Declaration on Principles of International Law concerning Friendly Relations and Co-operation among States in accordance with the Charter of the United Nations (Declaration on Friendly Relations)*.[134] A result of this duty is that '[n]o State or group of States has the right to intervene, directly or indirectly, for any reason whatever, in the internal or external affairs of any other State'.[135] Furthermore:[136]

> No State may use or encourage the use of economic, political or any other type of measures to coerce another State in order to obtain from it the subordination of the exercise of its sovereign rights and to secure from it advantages of any kind.

Similarly, and also arising from the *Declaration on Friendly Relations*, withdrawal of recognition from low-lying states actively engaged in securing their continued statehood would amount to a denial of peoples' right to self-determination.[137] It would breach the withdrawing state's 'duty to promote, through joint and separate action, realization of the principle of equal rights and self-determination of peoples'[138] and its 'duty to refrain from any forcible action which deprives peoples ... of their right to self-determination and freedom and independence'.[139]

---

[132]Montevideo Convention, art 3.

[133]Crawford (2006), p. 27.

[134]UNGA Declaration on Friendly Relations (1970), p. 123.

[135]Ibid.

[136]Ibid.

[137]Charter of the United Nations, art 1(2).

[138]UNGA Declaration on Friendly Relations (1970), pp. 123–124.

[139]Ibid., p. 124.

6.6 Legal Status of a Low-Lying State: Sovereignty

Continuity of statehood may be the best outcome for low-lying states to secure their rights and their capacity to safeguard their peoples' rights and interests as national citizens. However, there may be other means of maintaining some form of sovereignty, separate from statehood.

Whereas statehood refers to a particular status in international law, sovereignty defines the role of the state in the international legal system.[140] Crawford clarifies that sovereignty is an attribute of the state, not a precondition, and refers to the powers a state can exercise.[141] State sovereignty espouses two forms, internal and external sovereignty, the latter being explained in the *Island of Palmas* arbitration:[142]

> Sovereignty in the relations between States signifies independence. Independence in regard to a portion of the globe is the right to exercise therein, to the exclusion of any other State, the functions of a State. The development of the national organisation of States during the last few centuries and, as a corollary, the development of international law, have established this principle of the exclusive competence of the State in regard to its own territory in such a way as to make it the point of departure in settling most questions that concern international relations.

Crawford describes the two types of sovereignty as requiring government domestically and requiring independence internationally:[143]

> Government is treated as the exercise of authority with respect to persons and property within the territory of the state; whereas independence is treated as the exercise, or the right to exercise, such authority with respect to other states.

Whereas the concept of internal sovereignty is based around the ability of a state to make and enforce laws over its population within its territory, the international community has conceptualised this power over geographic space rather than people.[144]

External sovereignty focuses on the sovereign independence of states. Fowler and Bunck discuss how the term sovereignty is used to denote the independence of the state in the international community.[145] Sovereignty, as the concept is used in international law today, requires the upholding of two principles—sovereign equality and non-interference in the domestic affairs of a sovereign state. These principles originate in the 1648 Peace of Westphalia and are now codified in the UN Charter as principles of international law and the *Declaration on Friendly Rela-*

---

[140]Taylor (1997), p. 748.

[141]Crawford (2006), p. 32.

[142]*The Island of Palmas case* (1928), p. 838.

[143]Ibid., p. 55.

[144]Krasner (2001), p. 18.

[145]Fowler and Bunck (1995), p. 5.

tions.[146] Brownlie specifies that the traditional notion of sovereignty means legal personality of a particular kind, that of statehood.[147] Actions and roles most often associated with sovereignty include the ability to exercise exclusive jurisdiction over citizens, equality between states and the ability to enter into agreements with other states.

However, as most legal academics argue, sovereignty no longer applies to the state alone.[148] Sovereignty can be attributed to non-state entities without territory. While mentioning traditional views, Brownlie argues that the content of sovereignty can be conceptually viewed in different ways and relate to the legal competence of an entity.[149] A case in point is the sovereignty held by non-state sovereign entities of international law.

6.6.1 Non-State Sovereign Entities of International Law

Several non-state, *absque terra*, sovereign entities of international law exist on *sui generis* bases due to historical origins, such as the Sovereign Military Hospitaller Order of Saint John of Jerusalem, of Rhodes and of Malta (the Sovereign Order of Malta). Brownlie argues that, even if international lawyers should have regard to traditional principles, they cannot ignore the actuality that non-state entities do exist on the international plane.[150] Developing such an entity to succeed the low-lying states is mentioned briefly in the literature as an equitable solution. However, the entity is conceptualised as a new category of deterritorialised state.[151] Jessup has argued that a deterritorialised state is not possible as a state cannot be 'a kind of disembodied spirit'; it must have territory where the population resides and government exercises authority.[152] This conclusion is highly contestable.[153] Regardless, proposing a non-state sovereign entity—similar to those which already exist in international law—may provide a more politically palatable approach, especially for the partner state that works with the low-lying state on its relocation enterprise. As long as these sovereign entities are effective on the international legal plane and

---

[146]The Peace of Westphalia is a series of treaties signed in 1648. See Charter of the United Nations, art 2 and UNGA Declaration on Friendly Relations (1970).

[147]Brownlie (2008), p. 106.

[148]See Taylor (1997), Crawford (2006) and Brownlie (2008).

[149]Brownlie (2008), p. 119.

[150]Ibid., p. 64.

[151]Rayfuse (2011), p. 285.

[152]United Nations Security Council (1948), per United States Ambassador Jessup.

[153]The real examples, such as the Sovereign Order of Malta, show this conclusion to be incorrect. Other examples are indigenous communities, which often have many members who do not reside within a fixed region but maintain a close affinity and relationship with a central, tribal entity with governance functions. Whilst such realities create challenges for indigenous communities, it would be erroneous to describe the situation as 'a kind of disembodied spirit'.

do not breach any peremptory norm of international law,[154] and as long as there is recognition and consent by states, they constitute valid legal entities.[155] The recognition and consent should be achievable by referring to the historical reasons that underpin comparable entities, such as the Sovereign Order of Malta. The historical reasons for the low-lying states maintaining sovereignty are, first, their current existence as sovereign states. Secondly, the loss of habitable territory and permanent in situ population does not result from their own volition; it is due to the greenhouse gas emissions from other states that are arguably in contradiction with international legal obligations.

Another example of a non-state sovereign entity of international law is the Holy See. The UN classifies it as a non-member state, but Ryngaert considers it a sovereign non-state entity that the UN only labelled as a state because of confusion at the time on how to deal with the entity.[156] For his part, Brownlie doubts the Holy See is a state because it lacks a permanent population, and its sole purpose is religious.[157] This article follows the arguments advanced by Brownlie and Ryngaert that the Holy See is a sovereign, non-state entity. Brownlie describes how the Holy See lost its territory in 1870 and its first recognition as a sovereign entity occurred when Italy acknowledged in 1929 'the Sovereignty of the Holy See in the international domain'.[158] The Holy See's legal personality has since relied on recognition by other states. The Holy See has, among other things, been able to join inter-governmental organisations, become party to treaties, maintain diplomatic representation and have observer status at the UN.[159] The sovereignty allowing for such acts thus derives from its historic and spiritual sovereignty as the seat of the Catholic Church.[160] Overall, and in a practical way, the example of the Holy See underscores the view that an entity which was once a state but no longer meets the requirements of statehood can potentially become a sovereign entity of international law, with supporting recognition by the international community.

Because of the debate over the status of the Holy See, it is necessary to also consider the Sovereign Order of Malta. Cox explains that, historically, the Order fulfilled the requirements of statehood, but gained a different legal status after losing its territory, that of a non-state, sovereign entity of international law.[161] It began in the 11th century as a hospice in Jerusalem, and was recognised as an autonomous religious order dedicated to serve the poor and sick, later spreading

---

[154]Peremptory norms, also known as *jus cogens*, are principles of international law that are fundamental and no derogation is permitted: VCLT, art 53. Generally accepted peremptory norms include the prohibition of genocide, torture, and wars of aggression.

[155]Brownlie (2008), p. 64.

[156]Ryngaert (2011), p. 830.

[157]Brownlie (2008), p. 64.

[158]Ibid.

[159]Ryngaert (2011), p. 830.

[160]Ibid.

[161]Cox (2006), p. 212.

throughout Europe and gaining a military element.[162] At times, the Order exercised sovereign rule over Malta and Rhodes, but after losing its territory, the Order continued to run hospitals in Europe and maintain diplomatic relations with states.[163] Cox argues that the idea of the Order as a sovereign entity began when, in 1935, the Italian Court of Cassation held that it was so.[164] Today, any debate over the status of the Order in international law is minor.[165] The Sovereign Order of Malta maintains bilateral relations with 106 states that recognise it as a sovereign, non-state entity.[166] The Order is also a non-state permanent observer at the UN.[167] Furthermore, it has treaty-making power, is involved in active and passive legation, has membership of international organisations, and can mint stamps and issue passports (although these may not be universally recognised and complications can arise).[168] A similar sovereign status could be asserted by, what is still today, a low-lying state, to succeed that state. For example, for the purposes of the Vienna Convention on the Succession of States in respect of Treaties, the low-lying state would become the 'predecessor' and the new sovereign entity would become the 'successor State'.[169] No doubt, too, that the successor entity would need to work through succession details with its partner state.

6.6.2   Prospects of Forming Non-state Sovereign Entities

If low-lying states consider it desirable, the starting premise for new sovereign, non-state entities, like the Holy See and Sovereign Order of Malta, is their reliance on their universally-accepted statehood. However, the loss of their original territory materially altered their circumstances and, moreover, was due to no fault of their own. There are also religious reasons behind the sovereignty of the Holy See and the Sovereign Order of Malta, which distinguishes them from the low-lying states. This may be an argument against low-lying states creating other non-state sovereign entities of international law. However, there is nothing in international law that gives preference to theocracies over secular states (or vice versa). Indeed, it would be a bizarre development if the religious nature of the Holy See and the Sovereign

---

[162]Ibid., p. 214.

[163]Ibid.

[164]Ibid., p. 222.

[165]Brownlie (2008), p. 64.

[166]Sovereign Order of Malta (2016).

[167]United Nations (2016b).

[168]For example, a passport issued by the Sovereign Order of Malta would not be recognised in New Zealand under section 2 of the Passports Act 1992 due to inability to recognise nationality of holder.

[169]Vienna Convention on Succession of States in respect of Treaties 1946, art 2(1)(b)–(c).

Order of Malta constituted the *legal* basis (it would not only be a historical difference) for disqualifying low-lying states from using this mechanism.

Regardless of what distinguishes the religious orders from the low-lying states, nothing prevents new sovereign entities in international law from being created. Claude advances a pragmatic view, based on the ability of international law '[to avoid] fixed principles and strict norms and uniformly applicable rules, in favour of flexibility and discretion and improvisation – in sum, for pragmaticism as distinguished from principle'.[170] The former Secretary-General of the UN, Boutros Boutros-Ghali, promoted a similar notion in the UN context, stating that 'a major intellectual requirement of our time is to rethink the question of sovereignty ... to recognize that it may take more than one form and perform more than one function'.[171] There is, therefore, conceptual and legal space for the creation of non-state sovereign entities that may succeed the low-lying states.

While the transformation from state to non-state sovereign entity is plausible, it is important to remember that the development and making of international law depends on the consent of states.[172] The low-lying states possess their sovereign rights, in part, because of the universal recognition of their statehood by other states. Any limits on recognition of new sovereign entities could limit the rights currently enjoyed by the predecessor state.

6.7 Legal Status of a Low-Lying State: Legal Personality

If statehood and sovereignty were not available for a low-lying state, the possibility for them to retain some form of legal personality similar to that, for instance, of international organisations, could be envisaged. Legal personality is 'the capacity to be bearer of rights and duties under international law'.[173] In its advisory opinion, *Reparation for Injuries Suffered in the Service of the United Nations*, the ICJ acknowledged that the UN is an international legal subject, recognising the ability of non-state actors to have capacity and legal personality in international law.[174] Taylor argues that this shows that conceptions of the state as the centre of international governance are outmoded.[175] In reality, non-state actors are carrying out actions in regard to matters such as trade, human rights and the environment, which were all traditionally seen as functions of the state.[176] These types of activities will

---

[170]Claude (1993), p. 218.

[171]Boutros-Ghali (1992), pp. 98–99.

[172]Jennings and Watts (1992), pp. 125–126.

[173]G Schwarzenberger *A Manual of International Law* (6th ed, Security Council Official Records, London, 1976), as cited in Crawford (2006), p. 28.

[174]*Reparation for Injuries Suffered in the Service of the United Nations* [1949], pp. 178–179.

[175]Taylor (1997), p. 747.

[176]Ibid., pp. 746–747.

also be central concerns for the low-lying state, especially protection of the human rights of its nationals. Moreover, individuals and corporations can sometimes bring claims before international bodies established by treaties,[177] although states continue to ultimately control such fora and international legal personality is based on the consent of states.[178] Retaining legal personality would allow these low-lying states to invoke some rights and to perform duties on the international plane.

Nonetheless, such legal personality is clearly not equivalent to that of states. Some international organisations may ask the ICJ for advisory opinions, sign contracts with states, enter into treaties and attend UN meetings as observers. However, this status may not always offer privileges of sovereignty such as voting in international fora, full diplomatic protection, active legation, the ability to sign and uphold treaties, produce passports or have bilateral relationships with states. International personality will offer a low-lying state some privileges and an international voice, but will not offer as much as the status of a sovereign non-state entity or a sovereign state. This is particularly problematic, given the extent of risks to the citizens of low-lying states from the future that is not of their own making, and how those risks can be exacerbated or mitigated through developments of, and in, international law and fora.

6.8   No Legal Status for a Low-Lying State

The low-lying states may be left with no rights or standing under international law. This could occur voluntarily (by adopting migration solutions, for example), through coercion (by partner states not agreeing to host other sovereign entities on their territory), or through some mechanism developed for the purpose of extinguishing statehood. In other words, in these cases, the legal status of a low-lying state is neither statehood nor sovereignty nor legal personality. This is the least preferred option as the low-lying state will have neither the rights offered by these legal forms, nor their associated duties. It will have lost all of the rights it enjoys now as a full and sovereign state. Most importantly, the low-lying state will not be able to furnish protection to its citizens and will have to rely on other legal persons under international law to do so. Overall, the low-lying state will have no direct voice on the international plane, and—to reiterate—this will not be a consequence of their own actions, but other states' failure to mitigate greenhouse gas emissions and their subsequent failure to assist the low-lying states.

---

[177]Individuals generally lack standing to assert violation of international law in the absence of a protest by the state of nationality although states may agree to confer particular rights on individuals which will be enforceable under international law. See, for example, Optional Protocol to the International Covenant on Civil and Political Rights, art 1. For an example of a treaty conferring legal standing to corporations, see Convention on the Settlement of Investment Disputes between States and Nationals of Other States, art 25(2)(b).

[178]Crawford (2006), p. 29.

6.9 Dialogue Among States Now

In the current international legal order, low-lying states' futures are shrouded in ambiguity and uncertainty: after all, international legal norms did not develop with the physical ruination of a state's entire land territory in mind. Uncertainty as to the international community's reactions to this emerging scenario adds to the vulnerability of low-lying states and their populations. Not only are these states exposed to the physical effects of climate change, but they are also vulnerable to uncertain changes to their legal status in international law, and the associated uncertainties as to the fate of their national identities, cultures, traditions, and languages. The Prime Minister of Tuvalu, Epele Sopoaga, has recently called on the international community to urgently open a dialogue aiming at answering some of the legal questions posed, and to possibly consider novel, or adapt existing, legal norms directly addressing the scenario of state 'disappearance'.[179]

6.9.1 International Dialogue

As discussed earlier, international climate change themes have evolved to entail '[m]easures to enhance understanding, coordination and cooperation with regard to climate change induced displacement, migration and planned relocation, where appropriate, at the national, regional and international levels'.[180] Thus, the Conferences of the Parties to the UNFCCC would be a logical place for this relocation enterprise to develop.

Outside the UNFCCC process, the most universal forum for such a dialogue is the UN General Assembly. The General Assembly can 'discuss any questions or any matters within the scope of the present Charter or relating to the powers and functions of any organs provided for in the present Charter'.[181] The UN Charter also provides for the General Assembly to 'initiate studies and make recommendations for the purpose of . . . encouraging the progressive development of international law and its codification'.[182] (The International Law Association is currently examining this question of low-lying states' statehood, which would likely inform any General Assembly deliberations.) This chapter highlights a void in international law as regards the factual scenario confronting low-lying states and the failure of international law principles to provide requirements for statehood continuity. Comprising the vast majority of states, the UN General Assembly, although only vested with the power to pass non-binding resolutions, may be holding within reach some elements of adequate response. Although international law-making still rests mainly with states, the principles encapsulated in some UN resolutions could,

---

[179]Rowling (2016).

[180]UNFCCC (2010) Cancun Adaptation Framework, at [14(f)].

[181]Charter of the United Nations, art 10.

[182]Ibid., art 13(1).

depending on the wording and voting conditions, amount to customary international law and provide the starting point for a much needed debate on the legal fate of low-lying states.[183] The power of any such resolution would be heightened if it were to call upon existing international law, such as the sovereign equality of states, the right to self-determination, and the principles enshrined in the *Declaration on Friendly Relations*.

Alternatively, the UN General Assembly could, under article 96 of the UN Charter, request an advisory opinion from the ICJ on a legal question relating to climate change relocation. In the *Western Sahara* advisory opinion, the ICJ stated that it takes a substantive view of the term 'legal question'.[184] Furthermore, in the advisory opinion, *Accordance with law of the unilateral declaration of independence in respect of Kosovo*, a political aspect to a 'legal question' was held not to deprive the question of its legal character.[185] As the scenario contemplated in this article involves many legal issues, the ICJ could be asked for an advisory opinion. This could provide some guidance and help resolve some of the uncertainty. Although the ICJ has discretion to give advisory opinions, it has previously held in the *Legality of the Threat or Use of Nuclear Weapons* advisory opinion that it must have a 'compelling reason' for using its discretion *not* to give an opinion.[186] Advisory opinions are not binding, but they have significant impact on legal interpretation.

6.9.2 Regional Dialogue

Another option is to initiate regional dialogue. Low-lying states in the South Pacific are already discussing relocation solutions with neighbouring states.[187] Also, for the Maldives, dialogue has begun with India, Sri Lanka and Australia.[188] Regional discussion is important; the ICJ has indicated that a practice accepted by all the states in a region can amount to regional customary international law.[189] Thus, neighbouring states could adopt a common new practice backed by *opinio juris*, or enter into a binding treaty, thus establishing new international law norms that address the legal status of low-lying states. How states within the region interact with the low-lying state will be of the utmost importance as it is these neighbouring

---

[183]*Legality of the Use by a State of Nuclear Weapons in Armed Conflict* [1996], pp. 254–255.

[184]*Western Sahara* [1975], at [9]–[20].

[185]*Accordance with law of the unilateral declaration of independence in respect of Kosovo* [2010], at [27].

[186]*Legality of the Use by a State of Nuclear Weapons in Armed Conflict* [1996], p. 235.

[187]Bolatagici (2015).

[188]Ramesh (2008) and Doherty (2012).

[189]Regional customary international Law has been discussed by the International Court of Justice in the *Asylum Case* [1950], p. 276 and the *Right of Passage over Indian Territory Case* [1960], p. 39.

states which could allow the low-lying state—in whatever capacity—to be involved in regional fora.

The Pacific region is a case in point. The Pacific Islands Forum (PIF) could be used for dialogue on the ex situ continuity of states or other legal entities. Since 1971, the PIF has organised an annual meeting of heads of states and governments from the Pacific Island states, along with New Zealand and Australia.[190] The PIF promotes a collaborative approach to dealing with regional issues,[191] and climate change has been high on its agenda. The Pacific Islands Framework for Action on Climate Change 2006–2015 has had a broad scope—including mitigation, adaptation, governance, science, education, and cooperation—but it did not deal with the statehood question.[192] As a number of PIF members are susceptible low-lying states, dialogue on the emergence of regional norms and customary law principles on statehood is more likely to occur at the PIF than at the UN General Assembly.

Another grouping which could hold a dialogue on the continuity of low-lying states is AOSIS. AOSIS has been active in advocating for small island states at the UN, especially in relation to climate change. For example, at the second preparatory committee meeting for the UN Conference on Sustainable Development, speeches were given on behalf of AOSIS, both at the opening and closing sessions.[193] AOSIS could hold a summit similar to the AOSIS Climate Change Summit in 2009 and discuss the development of legal principles in regard to low-lying states and statehood.[194]

6.9.3 Purpose of Discussions

The objective of regional and international dialogue should be certainty by affirming the future legal status of low-lying states in a manner that, inter alia, respects their right of self-determination. A treaty—whether bilateral, sub-regional or regional—would offer the greatest certainty, since obligations therein must be upheld.[195] Law-making treaties can give rise to general norms for the future conduct of parties such as the norms created by the Hague Conventions of 1899 and 1907 (on the law of war and neutrality).[196] If a treaty was made with regard to low-lying states, it could lay out legal requirements for a state to retain statehood

---

[190]Barnett and Campbell (2010), pp. 102–103.

[191]This is encapsulated in the 'Pacific Way', an expression coined in 1970 by the Prime Minister of Fiji at the time, Ratu Sir Kamisese Mara, to explain the capacity of Pacific Islands to work together and to resolve their disputes through dialogue: see Costi and Boister (2006), p. 1.

[192]SPREP (2006).

[193]Thomson (2011) and Hussain (2011).

[194]UN-OHRLLS (2009).

[195]VCLT, art 26.

[196]Shaw (2008), p. 94; Brownlie (2008), p. 13.

(or other personality) and explain what the status of low-lying states can or will entail. The limitation here is that states not party to such a treaty would not be bound by its provisions.

Certainty may also come through the creation of customary international law. This requires general practice to be accepted as law.[197] Such practice is evidenced by 'diplomatic correspondence, policy statements, press releases, the opinions of legal advisers, official manuals on legal questions, international and national judicial decisions ... recitals in treaties and other legal instruments'.[198] No particular duration is required to establish a practice as long as there is consistency and generality of practice.[199] As a novel issue, if low-lying states are treated consistently and there is generality of practice, a new principle of customary international law could be formed relatively easily. Customary international law will also require belief by states that the principle is a legal obligation. There may be objectors to this custom but as mentioned, a custom could also be formed locally and this will indeed be more likely.

The dialogue could also go beyond discussing principles of international law and thus consider other issues such as the fate of the populations of the low-lying states and their rights to the maritime resources linked to its existing land territories. At the forefront of the dialogue, however, should be resolving the question of low-lying states' legal status, since most other related legal issues depend on the response to that question.

6.10 Conclusion

Nearly three decades have passed since the scenario confronting the low-lying states was first mentioned at the UN General Assembly. Even in the face of growing scientific consensus on climate change, the international community is yet to address the issue of low-lying states' legal status. There may be advantages for low-lying states in this situation, as it allows for individual solutions. They may want to avoid putting their future into the hands of the international community and whatever solution other parties 'invent' and impose upon low-lying states. After all, a collective resolution may well have undesirable aspects for a particular nation with particular circumstances. In any event, each low-lying state requires some sort of solution to this unprecedented and complex situation.

A key question shared by all of them relates to the future status of the legal person that is the state. A low-lying state could retain statehood, or become a non-state sovereign entity or another form of legal personality. The alternative is for each state to lose all their rights, privileges and status under international law,

---

[197]Statute of the International Court of Justice, art 38(1)(b).

[198]Brownlie (2008), p. 6.

[199]Ibid., p. 7.

and, therefore, the ability to protect their peoples and defend their interests. The potential for new and creative solutions that are in line with existing international law must be considered before answering too hastily this key question and assuming that statehood will be extinguished.

The Montevideo Convention codifies customary international law on the requirements for statehood. These include territory, population, government and capacity to enter into international relations. As this chapter has highlighted, there is the potential for all of these criteria to continue to be met *ex situ*, even though they cannot be sustained in situ. Couple the potential for creative solutions with the principle of state continuity and nothing suggests that statehood should dissolve.

An alternative to full statehood is sovereignty, which can be held separately from statehood. Non-state sovereign entities of international law do presently exist, offering a model for the low-lying states if full statehood is deemed undesirable or impossible. However, if sovereignty cannot be retained, the low-lying states should, at least, be able to maintain some legal personality. The range of entities with legal personality today is vast and there are clearly valid reasons for the low-lying states to justify legal personality.

Whilst there are no rules in international law that deal directly with the scenario facing the low-lying states, there are many existing international law principles that can support efforts to maintain national identity, culture, custom and tradition. A dialogue among states—bilaterally, sub-regionally, regionally, or internationally— is needed to actively consider this web of applicable international law, and if deemed desirable, forging new legal developments for this emerging, novel situation.

References

ABC News (2 April 2007) Kiribati's President: 'Our Lives Are At Stake'. ABC News. http://abcnews.go.com/WNT/story?id=3002001&page=1. Accessed 7 Sept 2016

Accordance with law of the unilateral declaration of independence in respect of Kosovo (Advisory Opinion) [2010] ICJ Rep 40

Administrative Court of Cologne (1989) *In re Duchy of Sealand* [1978] 80 ILR 683

Agreement between Tuvalu and Kiribati regarding their Maritime Boundary (29 August 2012)

Alliance of Small Island States, www.aosis.org

Anderson R (2004) A definition of peace. Peace Confl J Peace Psychol 10(2):101–116

Annan K (2004) The rule of law and transitional justice in conflict and post-conflict societies: Report of the Secretary-General (23 August 2004) S/2004/616

Asylum Case (Colombia/Peru) (Merits) [1950] ICJ Rep 266

Banderas News (24 June 2009) Mexico Claims to Gulf Shrink with Island's Loss. http://www.banderasnews.com/0906/nr-claimstogulf.htm. Accessed 7 Sept 2016

Barnett J, Adger WN (2003) Climate dangers and atoll counties. Climate Change 61:321–337

Barnett J, Campbell J (2010) Climate change and small Island states: power, knowledge and the South Pacific. Earthscan, London

Besson S (2009) The authority of international law – lifting the state Veil. Syd LR 31(3):343–380

Bolatagici L (1 December 2015) Fiji applauded for offer to take in people from Kiribati. The Fiji Times. http://www.fijitimes.com/story.aspx?id=332149

Boutros-Ghali B (1992) Empowering the United Nations. Foreign Aff 71(5):89–102

Brooks RE (2005) Failed states, or the state as failure? U Chi L Rev 72(4):1159–1196

Brownlie I (2008) Principles of international law, 7th edn. Oxford University Press, Oxford

Burkett M (2011) The nation *ex situ*: on climate change, deterritorialized nationhood and the post-climate era. Climate Law 2:354–374. doi:10.3233/CL-2011-040

Charter of the United Nations

Chimni BS (2007) The past, present and future of international law: a critical third world approach. Melb J Intl Law 8:499–515

Claude IL Jr (1993) The tension between principle and pragmatism in international relations. Rev Intl Stud 19(3):215–226

Convention on the Settlement of Investment Disputes between States and Nationals of Other States 575 UNTS 159 (opened for signature 18 March 1965, entered into force 14 October 1966)

Costi A, Boister N (2006) 'Régionalisation' du droit pénal international dans le Pacifique: enjeux et perspectives. In: Boister N, Costi A (eds) Droit Pénal International dans le Pacifique : Tentatives d'Harmonisation Régionale/Regionalising International Criminal Law in the Pacific. NZACL, Wellington

Cox N (2006) The continuing question of sovereignty and the sovereign military order of Jerusalem, of Rhodes and of Malta. Aust ILJ 13:211–232

Crawford J (1979) The creation of states in international law. Clarendon Press, Oxford

Crawford J (2006) The creation of states in international law, 2nd edn. Oxford University Press, Oxford

Crawford J (2012) Brownlie's principles of public international law, 8th edn. Oxford University Press, Oxford

Cuen D (11 September 2009) Mexico's Missing Island. BBC. http://www.bbc.co.uk/worldservice/documentaries/2009/09/090910_world_stories_mexico_missing_island.shtml.   Accessed 7 Sept 2016

Deposit by Kiribati of list of geographical coordinates of points, pursuant to article 16, paragraph 2, article 47, paragraph 9, and article 75, paragraph 2, of the Convention (2 January 2015) M.Z. N.111.2015.LOS

Deposit by the Republic of the Marshall Islands of geographical coordinates of points, pursuant to article 16, paragraph 2, article 47, paragraph 9, and article 75, paragraph 2, and article 84, paragraph 2 of the Convention (3 May 2016) M.Z.N.120.2016.LOS

Deposit by Tuvalu of lists of coordinates of points, pursuant to article 16, paragraph 2, article 47, paragraph 9, and article 75, paragraph 2, and article 84, paragraph 2, of the Convention (4 September 2013) M.Z.N.98.2013.LOS

Displacement Solutions (2010) Climate Change Displaced Persons and Housing, Land and Property Rights: Preliminary Strategies for Rights-Based Planning and Programme to Resolve Climate-Induced Displacement. Resource document. Displacement Solutions. http://displacementsolutions.org/files/documents/DS_Climate_change_strategies.pdf.   Accessed 7 Sept 2016

Doherty B (7 January 2012) Climate change castaways consider move to Australia. Sydney Morning Herald. http://www.smh.com.au/environment/climate-change/climate-change-castaways-consider-move-to-australia-20120106-1pobf.html. Accessed 7 Sept 2016

Dupont A, Pearman G (2006) Heating up the Planet: climate change and security, Lowy Institute Paper 12. Resource document. Lowy Institute for International Policy. http://www.lowyinstitute.org/publications/heating-planet-climate-change-and-security. Accessed 7 Sept 2016

Duursma JC (1996) Fragmentation and the international relations of micro-states: self-determination and statehood. Cambridge University Press, Cambridge

Expert Group on Global Climate Obligations (2015) Oslo principles on global climate change obligations. Resource Document. Global Justice Program. http://globaljustice.macmillan.yale.edu/news/oslo-principles-global-climate-change-obligations. Accessed 7 Sept 2016

Field CB, Barros VR, Dokken DJ, Mach KJ, Mastrandrea MD (eds) (2014a) Climate Change 2014: impacts, adaptation, and vulnerability: Working Group II Contribution to the Fifth Assessment Report of the Intergovernmental Panel on Climate Change. Cambridge University Press, Cambridge

Field CB, Barros VR, Dokken DJ, Mach KJ, Mastrandrea MD (eds) (2014b) Climate Change 2014: impacts, adaptation, and vulnerability: summary for policy makers. Cambridge University Press, Cambridge

Fitzpatrick M (2013) What accounts for the varying rates of sea level rise in different locations? Union of Concerned Scientists. http://www.ucsusa.org/publications/ask/2013/sea-level-rise.html. Accessed 7 Sept 2016

Flemming M (2014) Auschwitz, the Allies and Censorship of the Holocaust. Cambridge University Press, Cambridge

Foreign Policy (2016) Fragile States Index. http://foreignpolicy.com/fragile-states-index-2016-brexit-syria-refugee-europe-anti-migrant-boko-haram/. Accessed 7 Sept 2016

Foster JM (6 March 2014) Epic King Tides Offer Glimpse of Climate Change in Marshall Islands. Think Progress. https://thinkprogress.org/epic-king-tides-offer-glimpse-of-climate-change-in-marshall-islands-8a9d0d90947f#.aj5g4vsf3. Accessed 7 Sept 2016

Fowler MR, Bunck JM (1995) Power and the sovereign state. Pennsylvania State University Press, University Park

Gayoom MA (1987) Address to the United Nations General Assembly (42nd Session of the United Nations General Assembly on the Special Debate on Environment and Development, United Nations Headquarters, New York, 19 October 1987). http://maldivesmission.com/index.php?option=com_content&view=article&id=931:death-of-a-nation-speech&catid=93:statements-from-unga-42-1987&Itemid=56. Accessed 7 Sept 2016

Grant JP, Barker JC (2009) Parry & grant encyclopædic dictionary of international law, 3rd edn. Oxford University Press, Oxford

Grant TD (1999) Defining statehood: the Montevideo convention and its discontents. Colum J Transnatl Law 37:403–457

Hansen J, Sato M, Hearty P, Reudy R, Kelley M, Masson-Delmotte V, Russell G, Tselioudis G, Cao J, Rignot E, Velicogna I, Tormey B, Donovan B, Kandiano E, von Schuckmann K, Kharecha P, Legrande AN, Bauer M, Lo K (2015) Ice melt, sea level rise and superstorms: evidence from paleoclimate data, climate modeling, and modern observations that 2°C global warming is highly dangerous. Atmos Chem Phys 16:3761–3812

Harrison R (2007) The moral is: states make laws. In: Freeman M, Harrison R (eds) Law and philosophy: current legal issues, vol 10. Oxford University Press, Oxford

Hodgkinson D, Burton T, Anderson H, Young L (2010) 'The Hour When the Ship Comes In': a convention for persons displaced by climate change. Monash Univ Law Rev 36(1):69–120

Hussain T (2011) Statement by Ms Thilmeeza Hussain, Deputy Permanent Representative of the Maldives to the United Nations, on Behalf of the Alliance of Small Island States (AOSIS) During Closing Session of the Preparatory Committee Meeting of the UNCSD, 08 March 2011. United Nations Division for Sustainable Development. https://sustainabledevelopment.un.org/content/documents/18489aosis-closing.pdf. Accessed 7 Sept 2016

Immigration New Zealand (2016) Samoa Quota & Pacific Access Category. https://www.immigration.govt.nz/employ-migrants/hire-a-candidate/options-for-repeat-high-volume-hiring-new/samoan-quota-pacific-access-category. Accessed 7 Sept 2016

International Law Commission (2006) Draft articles on diplomatic protections with commentaries. YILC II(Pt 2):21–100

The Island of Palmas case (or Miangas) (The Netherlands v USA) (1928) 2 UN Rep Intl Arb Awards 829 [Island of Palmas case]

Jennings R, Watts A (eds) (1992) Oppenheim's international law, vol 1, 9th edn. Longman, London

Kahl C (2006) States, scarcity, and civil strife in the developing world. Princeton University Press, Princeton

King A, Henley BJ (17 August 2016) We have almost certainly blown the 1.5-degree global warming target. Australian Geographic. http://www.australiangeographic.com.au/topics/science-environment/2016/08/15-degree-global-warming-target-likely-to-be-surpassed. Accessed 7 Sept 2016

Klabbers J (2013) International law. Cambridge University Press, Cambridge

Krasner SD (2001) Rethinking the sovereign state model. Rev Int Stud 27:17–42

Legality of the Use by a State of Nuclear Weapons in Armed Conflict (Advisory Opinion) [1996] ICJ Rep 266

Legal Status of Eastern Greenland (Judgements, Orders and Advisory Opinions) (1933) PCIJ 53, Series A/B

Levine S (ed) (2009) Pacific ways: government and politics in the Pacific Islands. Victoria University Press, Wellington

Little A (15 December 2015) What the Paris climate agreement means for vulnerable nations. The New Yorker. http://www.newyorker.com/news/news-desk/what-the-paris-climate-agreement-means-for-vulnerable-nations. Accessed 7 Sept 2016

Maritime Delimitation and Territorial Questions between Qatar and Bahrain (Qatar v Bahrain) (Merits) [2001] ICJ Rep 40

Mathiesen K, Harvey F (8 December 2015) Climate coalition breaks cover in Paris to push for binding and ambitious deal. The Guardian. https://www.theguardian.com/environment/2015/dec/08/coalition-paris-push-for-binding-ambitious-climate-change-deal. Accessed 7 Sept 2016

McAdam J (2010) 'Disappearing States', Statelessness and the Boundaries of International Law. UNSW Law Research Paper No. 2010-2

McLeman R (2008) Climate change migration, refugee protection, and adaptive capacity building. McGill Int J Sustainable Dev Law Policy 4(1):1–18

Milman O (7 October 2015) UN drops plan to help move climate change affected people. The Guardian. https://www.theguardian.com/environment/2015/oct/07/un-drops-plan-to-create-group-to-relocate-climate-change-affected-people. Accessed 7 Sept 2016

Ministry of Environment, Energy and Water (2007) National Adaptation Program of Action: Republic of Maldives. Resource document. http://unfccc.int/resource/docs/napa/mdv01.pdf. Accessed 7 Sept 2016

Montevideo Convention on the Rights and Duties of States 165 LNTS 19 (opened for signature 26 December 1933, entered into force 26 December 1934) [Montevideo Convention]

New Zealand Government (2009) Report of New Zealand's Views on the Possible Security Implications of Climate Change. Report presented at the 64th UN General Assembly session, New York, September 2009

Office of the President of Kiribati (26 September 2009) President tells of greatest challenge. Press release. http://www.climate.gov.ki/2009/09/26/president-tong-addresses-the-un-general-assembly/. Accessed 7 Sept 2016

Office of the President of Kiribati (2010) Relocation. http://www.climate.gov.ki/category/action/relocation/. Accessed 7 Sept 2016

Optional Protocol to the International Covenant on Civil and Political Rights 999 UNTS 171 (opened for signature 16 December 1966, entered into force 23 March 1976)

Park S (2011) Climate Change and the Risk of Statelessness: The Situation of Low-lying Island States. Legal and Protection Policy Research Series, PPLA/2011/04. United Nations High Commissioner for Refugees

Paskal C (3 April 2010) Strange case of the disappearing islands. The New Zealand Herald. http://www.nzherald.co.nz/world/news/article.cfm?c_id=2&objectid=10635956. Accessed 7 Sept 2016

Peszke MA (2011) The British-Polish Agreement of August 1940: its antecedents, significance, and consequences. J Slavic Mil Stud 24:648–658. doi:10.1080/13518046.2011.624474

Powell JL (2016) Climate scientists virtually unanimous: anthropogenic global warming is true. Bull Sci Technol Soc 35(5–6):121–124. doi:10.1177/0270467616634958

Ramesh R (10 November 2008) Paradise almost lost: Maldives seek to buy a new homeland. The Guardian. https://www.theguardian.com/environment/2008/nov/10/maldives-climate-change. Accessed 7 Sept 2016

Randall A (7 October 2015) Not the end for displacement in Paris. UK Climate Change and Migration Coalition. http://climatemigration.org.uk/not-the-end-for-displacement-in-paris/. Accessed 7 Sept 2016

Rayfuse R (2009) W(h)ither Tuvalu? International Law and Disappearing States. UNSW Law Research Paper No. 2009-9

Rayfuse R (2011) International law and disappearing states: maritime zones and the criteria for statehood. Environ Policy Law 41(6):281–287

Reparation for Injuries Suffered in the Service of the United Nations (Advisory Opinion) [1949] ICJ Rep 174

Republic of the Marshall Islands Maritime Zones Declaration Act 2016

Respecting the Polish Forces in the United Kingdom AIR 2/4213 and WO 33/2389, National Archives, London

Right of Passage over Indian Territory Case (Portugal v India) (Merits) [1960] ICJ Rep 6

Rowling M (24 May 2016) Tuvalu PM urges new legal framework for climate migrants. Thomson Reuters Foundation News. http://www.reuters.com/article/us-humanitarian-summit-climatechange-mig-idUSKCN0YF2UD. Accessed 7 Sept 2016

Ryngaert C (2011) The legal status of the Holy See. Goettingen J Int Law 3:829–859. doi:10.3249/1868-1581-3-3-ryngaert

Ryngaert C, Sobrie S (2011) Recognition of states: international law or Realpolitik? LJIL 24 (2):467–490. doi:10.1017/S0922156511000100

Shaig A (2006) Climate change vulnerability and adaption assessment of the maldives land and beaches. Centre for Disaster Studies, Townsville

Shaw MN (2008) International law, 6th edn. Cambridge University Press, Cambridge

Shaw MN (2014) International law, 7th edn. Cambridge University Press, Cambridge

Shearer I (ed) (1994) Starke's international law, 11th edn. Butterworths, London

Sinclair, E. (2008) The Changing Climate of New Zealand's Security. Institute of Policy Studies, Working Paper 08/11, October 2008

Soanes C, Stevenson A (eds) (2009) Concise Oxford English Dictionary, 11th edn. Oxford University Press, Oxford

Sovereign Order of Malta (2016) Bilateral relations. https://www.orderofmalta.int/diplomatic-activities/bilateral-relations/. Accessed 7 Sept 2016

SPREP (2006) Pacific islands framework for action on climate change 2006–2015. Secretariat of the Pacific Regional Environment Programme. http://www.sprep.org/climate_change/pycc/documents/PIFACC.pdf. Accessed 7 Sept 2016

Statute of the International Court of Justice

Stocker TF, Qin D, Plattner G, Tignor MMB, Allen SK, Boschung J, Nauels A, Xia Y, Bex V, Midgley PM (eds) (2013) Climate Change 2013: The Physical Science Basis: Working Group 1 Contribution to the Fifth Assessment Report of the Intergovernmental Panel on Climate Change. Cambridge University Press, Cambridge

Taylor C (1997) A modest proposal: statehood and sovereignty in a global age. U Pa J Int Econ Law 18(3):745–809

Thomson P (2011) Statement by H.E. Mr. Peter Thomson, Ambassador and Permanent Representative of the Republic of the Fiji Islands to the United Nations, on Behalf of the Alliance of Small Island States (AOSIS) During the Opening Session of the Preparatory Committee Meeting of the Commission on Sustainable Development, United Nations Headquarters, New York, 07 March 2011. United Nations Division for Sustainable Development. https://sustainabledevelopment.un.org/rio20/2ndprepcom. Accessed 7 Sept 2016

Tinoco Claims Arbitration (United Kingdom v Costa Rica) (1923) 1 UN Rep Intl Arb Awards 369

UNFCCC (1992) United Nations Framework Convention on Climate Change 1771 UNTS 107 (opened for signature on 9 May 1992, entered into force 21 March 1994). [UNFCCC]

UNFCCC (2010) Report of the Conference of the Parties on its sixteenth session, held in Cancun from 29 November to 10 December 2010. FCCC/CP/2010/7/Add.1 (2010), at Decision 1/CP.16 [UNFCCC Cancun Adaptation Framework]

UNFCCC (2015a) Working of the Contact Group on item 3: Negotiating text: Advanced unedited version: 12 February 2015: Ad Hoc Working Group on the Durban Platform for Enhanced Action, Second session, part eight, 8–13 February 2015, Geneva, Switzerland

UNFCCC (2015b) Conference of the Parties: Twenty-first session: Paris, 30 November to 11 December 2015: Adoption of the Paris Agreement. FCCC/CP/L.9/Rev.1, annex. [UNFCCC Paris Agreement]

UNGA (1970) Declaration on Principles of International Law Concerning Friendly Relations and Co-operation Among States in Accordance with the Charter of the United Nations. GA Res 26/25, XXV (1970). [UNGA Declaration on Friendly Relations]

UNGA (2008) Follow-up to and Implementation of the Mauritius Strategy for the Further Implementation of the Programme of Action for the Sustainable Development of Small Island Developing States. GA Res 63/213, A/RES/63/213 (2008)

UNHCR (2009) Climate Change And Statelessness: An Overview, submission presented at 6th session of the Ad Hoc Working Group on Long-Term Cooperative Action under the UNFCCC, Bonn, 1–12 June 2009

UNHCR (2011) Summary of Deliberations on Climate Change and Displacement, April 2011. United Nations High Commissioner for Refugees. http://www.unhcr.org/4da2b5e19.pdf. Accessed 7 Sept 2016

UNGA (2012) Declaration of the High-level Meeting of the General Assembly on the Rule of Law at the National and International Levels. GA Res 27/1, A/RES/67/1 (30 November 2012)

UNHCR (2015) Guidance on Protecting People from Disasters and Environmental Change through Planned Relocation ("Guidance on Planned Relocation"). Brookings, UNHCR, and School of Foreign Service at Georgetown University

UN-OHRLLS (2009) The Alliance of Small Island States Climate Change Summit. United Nations Office of the High Representative for the Least Developed Countries, Landlocked Developing Countries and Small Island Developing States. http://unohrlls.org/meetings-con ferences-and-special-events/the-alliance-of-small-islands-states-aosis-climate-change-summit/. Accessed 7 Sept 2016

United Nations (2016a) UN Data. United Nations Data. www.data.un.org. Accessed 7 Sept 2016

United Nations (2016b) Non-member States. http://www.un.org/en/sections/member-states/non-member-states/index.html. Accessed 7 Sept 2016

United Nations Convention on the Law of the Sea 1833 UNTS 3 (opened for signature 10 December 1982, entered into force 16 November 1994)

United Nations Security Council (17 December 1948) Application of Israel for Admission to Membership in the United Nations, S/PV 383, S-III, per United States Ambassador Jessup. https://unispal.un.org/DPA/DPR/unispal.nsf/0/437DD877E349151B052566CE006D9189. Accessed 7 Sept 2016

Valemie R (22 March 2016) Climate Change. The Fiji Times Online. http://www.fijitimes.com/story.aspx?id=346536. Accessed 7 Sept 2016

Vienna Convention on Succession of States in respect of Treaties 1946 UNTS 3 (opened for signature 23 August 1978, entered into force 6 November 1996)

Vienna Convention on the Law of Treaties 1155 UNTS 331 (opened for signature 23 May 1969, entered into force 27 January 1980). [VCLT]

Wentz J, Burger M (2015) Designing a climate change displacement coordination facility: key issues for COP 21. Sabin Center for Climate Change Law, Columbia Law School, New York City

Western Sahara (Advisory opinion) [1975] 1975 ICJ Rep 12

Wikipedia. Solander Islands. https://en.wikipedia.org/wiki/Solander_Islands

Willcox S (2015) Climate Change Inundation and Atoll Island States: Implications for Human Rights, Self-Determination and Statehood. PhD thesis, London School of Economics and Political Science

World Commission on Environment and Development (1987) Report of the World Commission on Environment and Development: Our Common Future, 20 March 1987. United Nations. http://www.un-documents.net/our-common-future.pdf. Accessed 7 Sept 2016

Wyman KM (2013) Responses to climate migration. Harv Environ Law Rev 37(1):167–216

Yamamoto L, Esteban M (2014) Atoll Island states and international law: climate change displacement and sovereignty. Springer, Heidelberg. doi:10.1007/978-3-642-38186-7

Yamin F, Depledge J (2004) The international climate change regime: a guide to rules, institutions and procedures. Cambridge University Press, Cambridge

Chapter 7
Small States, Colonial Rule and Democracy

Derek O'Brien

7.1 Introduction

The Commonwealth Caribbean[1] is often singled out by scholars as one of the most democratic regions in the developing world.[2] The region is also notable for the number of small countries that it includes: whether measured by reference to population size or land mass, these are amongst the smallest countries in the world.[3] Based on the statistical link that has been drawn by scholars between small size and democracy,[4] it is tempting to conclude that the democratic character of the region is, therefore, largely a function of the small size of the majority of its countries.

Certainly, the consensus amongst scholars has been for many decades that 'small country size ...is conducive to democracy'.[5] Though it is always possible to

---

[1]The region comprises 12 independent countries that gained their independence from Britain at different times over a period of 20 odd years between 1962 and 1983: Antigua and Barbuda, Bahamas, Barbados, Belize, Dominica, Grenada, Guyana, Jamaica, St Kitts and Nevis, St Lucia, St Vincent and the Grenadines, and Trinidad and Tobago. It also includes a number of British Overseas Territories, but since they are not entirely self-governing they are not the focus of this chapter.

[2]Dominguez (1993), p. 57.

[3]For example, Barbados, Grenada, St Kitts and Nevis, and St Vincent and the Grenadines all have land mass area of less than 500 km^2; and, with the exception of Jamaica, which has a population of 2.7 million, Trinidad and Tobago, which has a population of just over 1.3 million, and Guyana which has a population of 800,000 the remainder of the countries within the region all have populations of under 400,000. Indeed, many of the countries are miniscule, with populations which hover around 100,000 or less: Antigua and Barbuda, Dominica, Grenada, St Kitts and Nevis, and St Vincent and the Grenadines.

[4]See, for example, Diamond and Tsalik (1999), pp. 117–60.

[5]As discussed in Srebnik (2004), pp. 329–341.

D. O'Brien (✉)
School of Law, Oxford Brookes University, Oxford, UK
e-mail: d.obrien@brookes.ac.uk

© Springer International Publishing AG 2017
P. Butler, C. Morris (eds.), *Small States in a Legal World*,
The World of Small States 1, DOI 10.1007/978-3-319-39366-7_7

identify small states where the record of democratic governance has been either inconsistent or non-existent—Cyprus, Fiji and Brunei, spring to mind—these have been regarded as very much the exception rather than the rule. Anckar and Anckar, thus argue that small size increases social cohesion and reduces the distance between citizens and their politicians.[6] Ott, too, suggests that because they are personalistic and informal, 'small-scale social structures encourage a more cooperative pattern of interaction among elites which is mimicked by the citizenry as a whole', and that 'size acts as an enabling environment for democratisation because the social system mitigates political conflict and increases the stake of citizens in the regime'.[7] Some scholars, such as Faris (1999), have even argued that island states are substantially more democratic than continental countries. This is, Faris argues, because their 'insulation' from the international system has allowed them 'to avoid getting embroiled in warfare and hence fostered a climate conducive to democratic politics'.

More recently, however, as scholars have had an opportunity to judge the effect of small size on democracy over a longer period of time they have begun to question the nature of the link. Some now argue that while there may be a correlation between small size and democracy this is not in and of itself evidence that it is the small size of these states that causes them to be democratic. Veenendaal, for example, argues that the maintenance of formal democratic institutions in small states has less to do with their size and more to do with their historic, geographical and political circumstances.[8] It is, he argues, no coincidence that most 'microstates' are former British colonies, which experienced longer and more intense periods of colonial rule.[9] Both the increased length and intensity of colonisation, according to Veenendaal, 'engendered a better socialisation in democratic values and traditions' among the population of these former colonies, creating a better environment after independence.[10] This view is supported by political scientists, writing about the Commonwealth Caribbean, who have argued that democracy flourished in the region because it had been 'socialized by over three hundred years of British colonialism', which had resulted in a 'deep penetration of the British influence'.[11]

In this chapter I wish to challenge both the 'historical circumstances' and 'small size' explanations of the democratic character of the region. In doing so, I wish to go even further and to challenge the very characterisation of the region as a bastion of democracy in so far as this is measured by reference to responsible and accountable government. I will begin, in Sect. 7.2, by contesting the proposition that 300 years of British colonial governance represented a prolonged tutelage in British democratic values, which served the region well as its political leaders took over the

---

[6]Anckar and Anckar (1995), pp. 211–229.

[7]Ott (2000), pp. 111–124.

[8]Veenendaal (2013), pp. 92–112.

[9]Ibid, p. 96.

[10]Ibid.

[11]Payne (1993), pp. 201–217.

reins of power. As I will show, colonial rule, or at least the system of 'Crown Colony' rule, which was the system of governance in force for most of the region for the last quarter of the nineteenth century and first half of the twentieth century, was characterised not by British democratic values but by the autocratic rule of a colonial Governor who had the final say on all matters affecting the colony concerned, and yet was neither accountable nor responsible to the people whom he governed.

In Sect. 7.3, I will draw, firstly, upon my analysis of the constitutional provisions surrounding the appointment and removal of key constitutional actors, which are common across the region; and, secondly, upon real life examples from countries across the region,[12] to demonstrate how the very small size of the majority of the countries in the region, combined with the extensive powers vested in the region's Prime Ministers, have enabled the latter to dominate almost every aspect of public life. As a result, it will be argued, periodic elections apart, Commonwealth Caribbean Prime Ministers are politically unaccountable in much the same way as were colonial Governors during the era of Crown Colony rule.

Finally, in Sect. 7.4, I will draw upon four very recent examples from St Kitts and Nevis and Antigua and Barbuda to show how the lack of any effective means of holding Prime Ministers politically to account meant that those who wished to challenge abuse of powers in these countries were obliged to invoke the Courts' powers of judicial review to prevent the Prime Minister, and other public officials and bodies subject to prime ministerial influence, from transgressing the limits imposed on them by the constitution. While there may be nothing inherently undemocratic in opponents of the government invoking the Courts' powers of constitutional review, neither is it the hallmark of responsible and accountable government.

In conclusion, I will argue that it is necessary in the light of the post-independence experience of the Commonwealth Caribbean not only to reconsider assumptions about the link between small size and democracy and about the 'civilising' effect of colonial rule, but also to re-evaluate the democratic credentials of a region which is often held up as a paradigm of postcolonial democracy.

7.2 Colonial Rule

The islands that make up what we now call the Commonwealth Caribbean came into British possession by different means and at different times over a period of nearly two and a half centuries, between 1624 and 1862. On the one hand, there were the *settled* colonies; so called because by British standards they had no 'civilised' inhabitants or settled law. Here the land, being 'desert and uncultivated',

---

[12]Trinidad and Tobago, Grenada, Antigua and Barbuda, Dominica, Jamaica, and St Kitts and Nevis.

to use Blackstone's words,[13] was claimed by the British by right of occupancy.[14] On the other hand, there were the *ceded* or *conquered* colonies; so called because the colony was acquired by conquest[15] or ceded to the British by another European power.[16] The means by which they came into British possession is important because it, typically, dictated the system under which they were governed.[17]

7.2.1   The 'Representative System'

The settled and ceded islands—Antigua, the Bahamas, Barbados, Dominica, Grenada, Jamaica, Nevis, St Kitts and St Vincent—were governed by what was known as the 'Representative System'. Under this system there were three branches of government: a Governor, representing the Crown, a Council, and a legislative assembly (the Assembly). The Governor, who was appointed by the King, had the power of granting or withholding his (they were all males) assent to any Bills which might be passed by the Assembly. The Council was composed of 'the most substantial men in the colony',[18] appointed by the King on the recommendation of the Governor, and could, therefore, be relied upon to support the Governor against the Assembly. The Assembly was comprised almost exclusively of white freeholders (mainly plantation owners) who were elected to represent the interests of the plantocracy.[19]

It was originally presumed that the Assembly would play a minor part in government. However, because any proposal for the expenditure of public money had to be approved by the Assembly, it possessed a powerful weapon which it was not afraid to use against the Governor whenever he sought to implement a policy which was deemed not to be compatible with the interests of its members and those whom they had been elected to represent—the plantocracy. It is arguable, therefore, that, in so far as the Assembly was able to hold the Governor to account by refusing to approve the expenditure of public money, the Representative System bore at least

---

[13]Blackstone (1765), p. 107.

[14]The settled colonies in the region comprised: St Kitts (1624), Barbados (1627), Nevis (1628), Antigua (1632), the Bahamas (1648), and Barbuda (1678).

[15]As in the case of Jamaica (1655), though it was treated by the British as if it were a settled colony, St Lucia (1762) and Trinidad (1797).

[16]As in the case of Dominica (ceded by the French in 1763), Grenada and St Vincent (ceded by the French in 1783), Tobago (ceded in 1793 by the French), Guyana (the territory, then know as Demerera, Berbice and Essequibo was purchased by the British in 1814, and in 1831 renamed British Guiana) and, finally, Belize (formally recognised as British following a treaty with Spain in 1763, and given the name of British Honduras in 1862).

[17]For a discussion of the distinction between settled colonies and ceded or conquered colonies see *Milirrpum v Nabalco Pty Ltd.* (1971) 17 FLR 141. See also Dupont (2001), p. 284.

[18]Wrong (1923), p. 40.

[19]Lewis (2004), p. 102.

some resemblance to the modern ideal of a government which is accountable to a body of elected representatives, even if those representatives were elected on an incredibly narrow franchise and were mainly concerned with protecting the interests of the plantocracy.

7.2.2 'Crown Colony' Rule

The conquered islands of Trinidad and St Lucia, by contrast, were governed from the outset by a system of 'Crown Colony' rule. Though the component parts differed in different colonies, the common elements of Crown Colony rule were a Governor, a Legislative Council and an Executive Council. The Governor sat as chairman of the Legislative Council, which was composed of an equal number of official members (senior civil servants appointed by the Secretary of State for the Colonies) and nominated unofficial members who were selected by the Governor, usually from among the dominant groups in each colony: the planter and merchant class. As Chairman, the Governor could always carry or veto any measure upon which the votes were evenly divided by virtue of his casting vote. The Governor also presided over the Executive Council, which had a purely advisory role, and was usually comprised of three ex officio members,[20] and two or more non-officials nominated by the Governor. The final say on matters affecting the colony thus always lay with the Governor who was responsible solely to the Secretary of State for the Colonies. Within the limits of their instructions from London Governors were, in effect, virtual autocrats.[21]

Though it was originally confined to conquered colonies, the system of Crown Colony rule became the mode of governance for the majority of countries in the region towards the final quarter of the nineteenth century as a result of the collapse of the Representative System following the 'Morant Bay rebellion', in Jamaica in 1865.[22] The rebellion and the brutal methods used to suppress it made it clear to the British authorities that the Representative System, which had been entirely geared towards protecting the interests of the plantocracy, was completely unsuited to the post-emancipation political landscape of these former slave colonies. Introduced to Jamaica in 1866, Crown Colony rule was subsequently extended to all the other countries in the region; with the exception of the Bahamas and Barbados, which retained the Representative System.

The extension of Crown Colony rule, which deprived the majority black population of any opportunity to participate in the government of their country, was opposed from the outset in Jamaica, and during the first quarter of the twentieth century hostility towards Crown Colony rule gathered pace across the region. This

---

[20]Lewis (2004), p. 98.

[21]Ibid.

[22]Holland Rose (1940), pp. 735–37.

was especially so in a country like Trinidad, which had not had a single elected representative in its Legislative Council since it was established in 1831.[23] In response to this groundswell of discontent various modifications to the system were implemented in the 1920s and 1930s. These comprised: the introduction of a number of elected unofficial members into the Legislative Council; a wider franchise; and lower qualifications for candidates wishing to be elected to the Legislative Council. It was not, however, until the end of World War II that the British Government began in earnest to dismantle the system of Crown Colony rule, which had by then persisted in some countries for over a century, and to lay the foundations for a system of responsible government based on the Westminster model.

7.2.3   Independence and the 'Westminster Model'

Though the introduction of responsible government took place at different times in different colonies, the broad outline of the process was similar in each.

Firstly, universal adult suffrage was introduced as the property and income qualifications were replaced by a simple literacy test, which itself was eventually abandoned. Secondly, the number of elected members in the Legislative Councils was incrementally increased, while the number of ex officio and nominated unofficial members was correspondingly decreased. In some cases, such as Jamaica and Trinidad and Tobago, the Legislative Council was replaced by a bicameral Parliament modelled on the British Parliament, with a wholly elected lower house (the House of Representatives) and a nominated upper house (the Senate), Thirdly, the Executive Council, formerly a purely advisory body, became the principal policy-making body. At the same time, the number of members drawn from the elected element of the Legislative Council steadily increased until it reached the point when the elected members formed a majority on the Executive Council. Fourthly, the semblance of Cabinet government began to emerge as one of the elected members of the Executive Council was appointed as Chief Minister with the approval of the House of Representatives, which also had the power to dismiss the Chief Minister by majority vote, and the Governor assigned portfolios to the other elected members of the Executive Council on the Chief Minister's recommendation. In this way the autocratic rule of the Governor under the system of Crown Colony rule was gradually displaced by what was supposed to be a system of collective, democratic self-government.[24]

All of this, however, took place in a very concentrated period, which meant that by the time of independence most countries in the region had barely a decade's worth of experience of functioning under a responsible and accountable system of

---

[23]Lewis (2004).

[24]Meighoo and Jamadar (2008).

government. As we have seen, for almost a century before that, and in some cases for more than a century, they had experienced only autocratic rule under a colonial Governor. While it is true that a number of the region's independence leaders and politicians had been educated in England and were familiar with the Westminster model, they nevertheless represented a very small section of the wider political class in the region.[25]

The transplantation of the Westminster model to the Commonwealth Caribbean, where the experience of colonial rule had been quite different to that of the 'white' Dominions of Canada, Australia and New Zealand, was, therefore, something of an experiment. Certainly, there was very little in the region's history to suggest that this transplantation would successfully take root. Though it was relatively straightforward to replicate the institutions upon which the Westminster model is based—the Queen as head of state, a Cabinet, Parliament, the Courts and a civil service—it was always going to be difficult to replicate the culture of conventions, habits and understandings which underpin relations between these institutions. This culture had been gradually evolving in Britain over a number of centuries, ever since the Glorious Revolution of 1688, and was the end product of a quite different set of political circumstances and needs. A history of slavery and a century of Crown Colony rule, by contrast, was hardly the most fertile soil in which to attempt to transplant the Westminster model. As that doyen of West Indies studies, GK Lewis, observed of Crown Colony rule: 'as a system...it robbed all who participated in it of self-respect',[26] echoing the observation of the Jamaican independence leader, Norman Manley, that: 'the system...was a perfect instrument for the degradation of political life'.[27] Its lasting imprint on the political culture of the region can be seen in the next section as we examine the emergence in the post-independence era of very powerful Prime Ministers, dominating almost every aspect of public life.

7.3 Powers of Commonwealth Caribbean Prime Ministers

In accordance with the Westminster model of government,[28] each of the region's constitutions vests 'executive authority' in the head of state—in most cases the Queen acting through her representative, the Governor General. 'Executive power',

---

[25]Norman Manley, Jamaica's first Prime Minster, had served in the Royal Field Artillery in World War I and was a Rhodes scholar at Oxford; Eric Williams, Trinidad and Tobago's first Prime Minister competed his PhD at Oxford; Errol Barrow, the first Prime Minister of Barbados concurrently studied law at the Inns of Court and economics at LSE; and Forbes Burnham, the first Prime Minister of Guyana also attended LSE where he studied law.

[26]Lewis (2004), p. 101.

[27]Quoted in ibid.

[28]With exception of Guyana, which in 1980 adopted a Cooperative Socialist Republic Constitution, all of the other countries in the region have been governed under the Westminster model since independence.

on the other hand, is vested in the Cabinet, comprising the Prime Minister and such other ministers from among the members of the legislature as the Prime Minister selects. The Cabinet is thus collectively charged with the general direction and control of government: deciding issues of policy, both domestic and foreign, and how public money should and should not be spent.

Typically under the Westminster model, the Prime Minister, as head of the Cabinet, automatically wields enormous political power by virtue of his or her right of proposal and veto; to appoint and delegate responsibilities to ministers and departments; to be consulted about all significant matters relating to government policy; and to set the government's policy agenda. The Prime Minister will, however, be even more powerful within his or her own party and within Cabinet where he or she is electorally very successful. This happens with especial regularity in the Commonwealth Caribbean. This is in no small part due to the 'first-past-the-post' electoral system, which is the electoral system of choice across the region,[29] and which historically has tended to result in a disproportionate majority for the winning party in terms of seats won and votes cast. This makes it easier for Prime Ministers to win electoral landslides,[30] and renders them politically unassailable, at least from within their own party.

The Prime Minister's paramountcy both within Cabinet and within their own party makes it even more important that there exists a wider network of institutions and public officials who are charged by the constitution to provide a check on the Prime Minister, to guarantee the neutrality of the public service, to ensure the integrity of the electoral process, and to supervise the expenditure of public finances. As we will see below, however, in the Commonwealth Caribbean the Prime Minister's involvement in the appointment of the key officials in these bodies makes it difficult for these officials to fulfil the constitutional role allocated to them.

7.3.1 Powers of Appointment and Dismissal

Commonwealth Caribbean constitutions vest Prime Ministers with very extensive powers of appointment. Indeed, it would be no exaggeration to say that there are hardly any senior public officials who do not owe their appointment either directly or indirectly to the Prime Minister.[31]

---

[29]With the exception of Guyana which has a party list system.

[30]For example, in Grenada in 1999 when the New National Party under Keith Mitchell won all of the available seats; in St Lucia in 1997 and 2001 when the St Lucia Labour Party under Kenny Anthony won 16 and 14 out of the 17 available seats; Barbados in 1999 when the Barbados Labour Party under Owen Arthur won 26 of the 28 available seats; and in St Kitts and Nevis in 2000 and 2004 when the St Kitts and Nevis Labour Party under Denzil Douglas won 8 and 7 out of the 11 available seats respectively. See Barrow-Giles and Joseph (2006), pp. 5–6.

[31]In addition to those identified in this section the Prime Minister appoints ambassadors, high commissioners and other principal representatives of the state. See, for example, Antigua and Barbuda Constitutional Order 1981 (Constitution of Antigua), s 101(2)(c).

7.3.1.1 Governor General

For those countries in the region that remain constitutional monarchies—that is every country with the exception of Dominica, Guyana and Trinidad and Tobago[32]—the head of state is the Queen, who acts at all times through her appointed representative in the country concerned, the Governor General. This means that whilst the titular head of state is the Queen, for all practical intents and purposes the effective head of state is the Governor General.[33]

Though it is not mentioned in the text of any of the region's constitutions, Governors General are customarily appointed in accordance with the recommendation of the Prime Minister of the country concerned.[34] There is no requirement that in making their recommendation the Prime Minister must consult with any other person or body. No specific qualifications are required for appointment as Governor General and a candidate's previous political affiliations are no disqualification. Indeed, a number of Governors General within the region have previously been members of the same political party as the incumbent Prime Minister and have even held high political office.[35] The custom of appointment upon the advice of the Prime Minister is taken also to extend to the Governor General's dismissal.[36] Governors General are thus removable at the request of the Prime Minister and there have been several instances, since independence, of the premature dismissal or resignation of Governors General.[37]

The contingent status of Governors General is not unique to the Commonwealth Caribbean: it also applies to Governors General elsewhere in the Commonwealth. However, the political as well as social ties between the Governor General and the Prime Minister are usually much closer in the tightly knit communities that are so characteristic of the small islands of the Caribbean than is the norm elsewhere; with the Prime Minister and Governor General often having attended the same schools, the same churches and the same clubs. When contingent status is combined with a

---

[32]Dominica embarked upon independence as a republic with a ceremonial President. Guyana became a republic in 1970 when it replaced the Queen as head of state with a ceremonial President and Trinidad and Tobago followed suit in 1976.

[33]Once appointed, the Governor General is a free agent who does not need to receive instructions from the Queen. See Kumarasingham (2010), p. 45.

[34]Dale (1983), p. 112.

[35]O'Brien (2014): This is true, for example, of all the Governors General in the Bahamas since independence, with one exception. It is also true of Sir Deighton Ward in Barbados; Carlyle Glean in Grenada; Clifford Campbell, Sir Florizel Glasspole and Howard Cooke in Jamaica; and Allen Lewis, Boswell Williams and George Mallett in St Lucia. Antigua had until recently been an exception to this general trend. However, the most recently appointed Governor General in that country, not only contributed to the governing party's election campaign, but also made frequent public appearances to support the party's election campaign.

[36]Dale (1983), p. 113. See also Bogdanor and Marshall (1996) who have, in contrast, suggested that the Queen may retain a residual discretion to refuse a Prime Minister's request for dismissal of a Governor General.

[37]O'Brien (2014), p. 48.

close political and social nexus it creates a highly toxic mix, which can impact upon the constitutional role of the Governor General in two ways. Firstly, it can undermine the Governor General's independence when called upon to exercise their so-called 'reserve powers': such as the power to summon, prorogue or dissolve Parliament and to appoint or dismiss the Prime Minister.[38] Thus, for example, when deciding whether or not to grant a Prime Minister's request for the dissolution of Parliament it may be difficult for the Governor General, in those countries where they are empowered to do so,[39] to base their decision on what is in the best interests of the country rather than what is most politically expedient for the Prime Minister. Secondly, the close relationship between Governors General and Prime Ministers can undermine the former's political neutrality when exercising their power, for example, to appoint independent senators. In such cases the Governor General is expected to act in their own deliberate judgement, but their proximity to the Prime Minister means they may be reluctant to appoint persons who are likely to be critical of the government.[40]

7.3.1.2   Ministers

Under the Westminster model it is customary for members of the Cabinet to be appointed and removed by the head of state upon the recommendation of the Prime Minister. However, the appointment of government ministers upon the recommendation of the Prime Minister in the Commonwealth Caribbean is distinctive in at least two respects, both of which have implications for the legislature's willingness and ability to act as a check upon the Prime Minister.

Firstly, there is no rule about the number of elected members whom the Prime Minister can appoint to his Cabinet. Prime Ministers can, therefore, pack the government benches in Parliament with members of his or her own Cabinet, who are bound by the convention of collective responsibility to toe the government line or resign their post as minister.[41] Indeed, in the very small Parliaments of the Eastern Caribbean, which have fewer than 20 members, it is not unusual for all the elected members from the governing party to be government ministers.[42]

---

[38]The reserve powers of Governors General also include, by implication, the power to refuse assent to a Bill presented by Parliament, though by a convention which applies as much to the Caribbean as it does elsewhere in the Commonwealth, a Head of State under the Westminster model must always assent to a Bill on the advice of Cabinet. Indeed, were a Governor General to refuse to accept the Cabinet's advice assent to a Bill presented by Parliament, it would, save in the most exceptional circumstances as where the Bill abolished the independence of the judiciary, be regarded as profoundly undemocratic. In the case of this particular reserve power then the contingent status of the Governor General is not regarded as being problematic.

[39]Belize, St Lucia and St Vincent and the Grenadines.

[40]Robinson et al. (2015), p. 96.

[41]In Jamaica, for example, under the Prime Minister PJ Patterson 22 out of the 34 elected members of the House of Representatives were government ministers.

[42]Robinson et al. (2015), p.103.

Secondly, there is no rule preventing the Prime Minister from appointing as a senator a candidate who has been rejected by the electorate and, once they have been appointed as a senator, recommending their appointment as a government minister.[43] This occurred in Trinidad and Tobago, following the 2000 election, when Prime Minister Panday requested that the President appoint as senators seven of his party's candidates who had just lost their seats in the election. His request was initially refused by the President, who considered that 'using people who have been, to put it in this way, rejected by the electorate in a representative and democratic system' in such large numbers was 'unprecedented'.[44] However, a compromise was eventually agreed, with the President agreeing to appoint the seven senators upon the Prime Minister undertaking that only two would be recommended for appointment as members of his Cabinet. The practice of appointing defeated electoral candidates as senators who go on to become ministers is not, however, confined to Trinidad and Tobago. Following the 1995 elections in Grenada, seven out of the 13 senators who were appointed on the recommendation of the Prime Minister were defeated candidates in the general election, and five of these defeated candidates went on to be appointed government ministers.[45]

The presence of a significant number of government ministers, who are bound to toe the government line, makes it more difficult for senates to perform their constitutional function of serving as a check on the executive. However, as we will see below, the appointment of government ministers as senators is not the only means by which Prime Ministers can exert their influence over that body.

7.3.1.3 Senators

With the exception of Belize,[46] a majority of the senators within each legislature are appointed for a fixed period of 5 years (corresponding to the lifetime of a Parliament) upon the recommendation of the Prime Minister and may also be removed at any point during this five-year period upon the recommendation of the Prime Minister.[47] This means that, save in cases where the Prime Minister is attempting to drive through reforms to the constitution (which, typically, require a special two thirds or three quarter majority of both Houses of Parliament), the nominated element in Commonwealth Caribbean legislatures rarely acts as an effective check on the Prime Minister. Senators who have the temerity to oppose

---

[43]Though in some countries there is a rule about the number of senators that may be appointed as ministers. See, for example, Jamaica (Constitution) Order in Council 1962 (Constitution of Jamaica), s 69(3).

[44]Ghany (2002).

[45]Ibid.

[46]Belize Constitution (Sixth Amendment) Act 2008.

[47]Dominica, St Kitts and Nevis and St Vincent and the Grenadines have unicameral legislatures, which include elected members and nominated senators. Nominated senators are not entitled to vote on a motion of no confidence.

the Prime Minister's wishes must face the risk of having their appointment summarily revoked. In Trinidad and Tobago, for example, in 2000, the President was obliged, upon the recommendation of the Prime Minister, to revoke the appointment of two senators who had voted against legislation proposed by the Government.

7.3.1.4   The Office of Speaker

Speakers of the region's Lower Houses are elected by a majority of the members of the Lower House. They are expected to function in the same way as their counterpart in the British Parliament, but are not subject to the same set of conventions, which are designed to emphasise the Speaker's political neutrality. For example, there is no convention that a retiring Speaker will be replaced by someone from the Opposition. In the absence of such conventions, regulating the Speaker's election and their conduct while in office, it has proved to be extremely difficult in the Commonwealth Caribbean to establish a Speakership, which is seen to be impartial and which can earn the respect and trust of all political parties. Instead, the tendency across the region has been for the office of Speaker to be regarded as the privilege of the party in power, making it almost impossible to disassociate the Speaker from party politics.

In Antigua and Barbuda, for example, there was for over a decade a protracted dispute between the Antigua and Barbdua Labour Party (ABLP) and the Speaker of the House of Representatives, whom the ABLP believed had consistently favoured the governing party in her management of business in the House of Representatives. In the course of this dispute Gaston Browne, an ABLP member of the House of Representatives, unsuccessfully sought to challenge the legality of the Speaker's appointment on the ground that she was disqualified because she was the holder of a public office, namely Executive Secretary to the Board of Education.[48] Though it is impossible to establish whether this was by coincidence or not, the day after his legal challenge to the Speaker's appointment was dismissed by the High Court[49] Browne was suspended indefinitely from Parliament by the Speaker for protesting too vocally about the legitimacy of the Government.

7.3.1.5   Public Service Commissions (PSCs)

Under the Westminster model the traditional view of the role of public servants is based on principles that were first outlined in the Northcote-Trevelyan report of 1854, the essential characteristics of which are: a permanent bureaucracy staffed by

---

[48]Contrary to the Constitution of Antigua, s 39(1)(g).

[49]*Browne v Giselle Isaac-Arrindell*, High Court Antigua, 16 June 2010. Unreported. Available on file with the author.

neutral and anonymous officials; recruitment and promotion based on merit; self sufficiency; and a strict separation of power between government ministers who decide policy and the public servants who administer it.

In support of these principles the constitutions of all the countries within the region make provision for the establishment of a PSC, which is supposed to have exclusive powers over the appointment, promotion, transfer and removal of public servants (which term includes members of the Civil Service, the Teaching Service and Police Service). To secure their independence each constitution imposes strict conditions on eligibility for membership of a PSC,[50] length of appointment[51] and security of tenure.[52] There is also usually a provision that the conditions of service of PSC members shall not be altered to their disadvantage. In this way it was hoped that public servants would be insulated from political interference by the government of the day.[53]

Notwithstanding these prophylactic devices, appointments to PSCs throughout the region continue to remain very much within the Prime Minister's sphere of influence. In a number of cases members of the PSC, including the Chairman, are appointed by the head of state only after seeking the advice of the Prime Minister.[54] Though the Prime Minister is usually required to consult with the Leader of the Opposition before tendering advice to the Governor General, in most cases consultation is no more than a formality, since the Prime Minister is not required to obtain the Leader of the Opposition's agreement to his or her preferred candidate.[55] As Basdeo Panday, the former Prime Minister of Trinidad and Tobago, caustically observed when describing the process of consultation when he was the Leader of the Opposition: 'I am consulted by a letter written by the President's secretary to me saying that they are going to appoint so and so and if I have any comments or objections. That is the level of consultation'.[56] As a result the risk of indirect prime

---

[50]Thus, former public officers and members of the legislature are usually disqualified (see, for example, Saint Christopher and Nevis Constitution Order 1983 (Constitution of St Kitts and Nevis), s 77(2)) and there is also usually a quarantine period during which a person who has held office or acted as a member of a PSC cannot be eligible for appointment to another public office (see, for example, Constitution of the Republic of Trinidad and Tobago Act (Constitution of Trinidad and Tobago), s 126(2).

[51]Appointments to the PSC are for a fixed term, the minimum being 2 years, as in the case of Antigua, and the maximum being 5 years, as in the case of Jamaica.

[52]Members of the PSC may only be removed from office by the President or Governor General, as the case may be, for inability to discharge the functions of their office whether arising from infirmity of mind or body or any other cause or for misbehaviour, and then only if their removal has been recommended by a tribunal, comprising a chairman and two other members appointed by the Chief Justice (see, for example, Constitution of St Kitts and Nevis, ss 77(5) and (6).

[53]*Thomas v AG Trinidad* [1982] AC 113.

[54]Constitution of Antigua, s 99(1).

[55]See, for example, Constitution of Antigua, s 99(1); Constitution of Trinidad and Tobago, s 120 (1).

[56]O'Brien (2014), p. 166.

ministerial influence over the functions of the PSC remains a profound concern in the region. George Eaton, for example, writing about the operation of the PSC in Grenada, has argued that:

> It is naïve to suppose that persons appointed through a process of political patronage to a legally independent, but functionally very important body, will thereafter become immune to political sensitivities ... It simply means that political persuasion is exercised surreptitiously or, as in the case of Grenada, under the Gairy regime, by the outright usurpation and subversion of the functions and responsibilities of the [PSC].[57]

Thus at each stage of a public servant's career, the possibility exists of interference in that career by a politically directed PSC.

7.3.1.6 Election Management Bodies and Constituency Boundaries Commissions

Responsibility for the management and administration of elections is assigned by each of the region's constitutions either to a Supervisor of Elections/Parliamentary Commissioner or to an Electoral Commission.[58]

Where responsibility is assigned exclusively to a Supervisor of Elections/Parliamentary Commissioner,[59] appointed by either the PSC or the Governor General,[60] the concern is that both of these appointing bodies are susceptible to the influence of the Prime Minister for the reasons outlined above.[61] Similar concerns have been expressed about potential political interference with Electoral Commissions in the region. Usually, the majority of members of Electoral Commissions are appointed on the Prime Minister's recommendation.[62] Though the Chairman of the Election Commission is usually appointed by the Governor General 'acting in his own deliberate judgment',[63] for the reasons discussed above it is unlikely that the Governor General would appoint as Chairman anybody who was not first approved by the Prime Minister. Even where there is a constitutional requirement that the Governor General must act on the recommendation of the Leader of the Opposition when appointing a proportion of the members of the Electoral Commission, the members so appointed will always form a minority. Moreover, there have been a number of occasions where there has been no Leader of the Opposition to consult

---

[57]Eaton et al. (2002), p. 210.

[58]With the exception of Antigua and St Kitts, where responsibility is shared, somewhat uncomfortably, between the Supervisor of Elections and the Electoral Commission.

[59]As in the Bahamas, St Vincent and the Grenadines, and Grenada.

[60]See, for example, Grenada Constitution Order 1973, s 35 (1) which provides that the Supervisor of Elections should be appointed by the Governor General.

[61]See Antigua Constitution Review Commission Report (2002), p. 89.

[62]As in Barbados and Belize, where three of the five members are appointed in accordance with the advice of the Prime Minister.

[63]See Constitution of Saint Lucia 1979, s 53(3).

either because the Opposition has boycotted the general election, as in the case of Trinidad and Tobago in 1971 and Jamaica in 1983, or the governing party has won all of the available seats in the legislature.

The other key electoral body within the Commonwealth Caribbean are Constituency Boundaries Commissions (CBCs), which have the task of reviewing and making recommendations to Parliament about the size and number of constituencies in each country.[64] In recognition of the importance of the task assigned to CBCs there are a number of express provisions within each Constitution designed to ensure a measure of political neutrality as well as political balance in the composition of each CBC. The degree of neutrality and balance thus achieved is, however, questionable. For example, in the case of the Bahamas, Dominica, Grenada and St Lucia the Chairman is, ex officio, the Speaker of the House of Representatives. As we have seen, the Speaker is elected by the government majority in Parliament, thereby tending to undercut any claim to political neutrality. As the Eastern Caribbean Court of Appeal noted in *Constituency Boundaries Commission and Another v Baron*[65] with regard to those countries where the Chairman of the CBC is, ex officio, the Speaker:

> The reality of the situation is that when such a Commission is being set up, the respective sides will recommend members whom they are satisfied will look after each side's respective interests. Their concentration will be more on political advantage than constitutional requirements. I agree that the Speaker as Commission Chairman stands in the middle. But again one has to be real. The Speaker was elected by a government majority.

In the case of Antigua and Barbuda and St Kitts and Nevis, the Chairman is even more obviously a political appointee; being appointed in each case by the Governor General, acting in accordance with the advice of the Prime Minister after the latter has consulted the Leader of the Opposition.[66]

Being able to exert influence over the Chairman of the CBC is important to Prime Ministers because of the Chairman's casting vote in the event of a tie. This is crucial in those countries where members of the CBC are appointed in equal numbers upon the recommendation of the Prime Minister and the Leader of the Opposition.[67] By virtue of their casting vote there is always the potential that a

---

[64]In Barbados and Trinidad, the Electoral Commission and Boundaries Commission are combined in one body—the Electoral and Boundaries Commission. The exception is Jamaica, where under s 67 of the Constitution of Jamaica, this responsibility is assigned to a Standing Committee of the House of Representatives.

[65][2001] 1 LRC 25.

[66]Constitution of Antigua, s 63(1)(a). See also the Constitution of St Kitts and Nevis, s 49(1)(a): which notes that the Chairman is appointed by the Governor General, acting in accordance with the advice of the Prime Minister given after the Governor General has consulted the Leader of the Opposition and such other persons as the Governor General, acting in their own deliberate judgement, has seen fit to consult.

[67]In Antigua, there is an even greater imbalance as two members are appointed by the Governor General in accordance with the advice of Prime Minister and only one member in accordance with the advice of Leader of the Opposition, thus affording the Prime Minister the final say in the

Chairman who is susceptible to the Prime Minister's influence will ensure that the government's view prevails in the event of a tie when the CBC is taking decisions about the review of constituency boundaries.

7.3.1.7 Auditor General

Auditor Generals in the region are appointed by the President or Governor General, as the case may be, acting in accordance with the recommendation of the PSC after the latter has informed the Prime Minister of their recommendation.[68] The involvement of the Prime Minister in this process, even if it is indirect, is highly problematic because of the nature of the constitutional role of Auditors General. This entails auditing and reporting on the public accounts of all government departments and serving as a 'watchdog' on behalf of the public to guard against any impropriety in the conduct of the public finances. As a result, it is almost inevitable that at some point Auditors General will be brought into conflict with the government of the day. This can be most clearly seen in the case of the Auditor General of Grenada, who was removed from office for a letter of rebuke which she had written to the Prime Minister after she had discovered not only that the report that she had sent him to lay before Parliament had been tampered with, but also that there had been inordinate delay on the part of the Prime Minister in laying her report before Parliament.[69]

Because of the potential for conflict with the executive the safer course for Auditors General who wish to remain in office is to do as little as possible. As the High Court of Antigua and Barbuda noted in the case of *Thomas v Harris*,[70] the failure of the Auditor General to file the audited accounts for 1985 to 1988 until 1999 and the accounts for 1989 until 2000, and the Prime Minister's failure to reprimand the Auditor General for his dereliction of his duties or to take any action to remedy the default, led to the inescapable conclusion that this state of affairs actually suited the Government, which had been in power since Antigua and Barbuda had been granted independence in 1981. In the Court's view, as a consequence of its longevity, the Government had 'perhaps grown complacent and unmindful of its public accountability to the citizens who had elected it in the first place'.

---

appointment of both the chairman and the majority of the members of the CBC (Constitution of Antigua, s 63(1). There is a similar imbalance in the appointed element of Bahamas CBC, but here at least the addition of two ex officio members—the Speaker as chairman and a Justice of the Supreme Court as deputy chairman—does mean the government is not guaranteed a majority on the CBC (Bahamas Independence Order 1973, s 69).

[68] Also known in Dominica, Grenada and St Kitts as the Director of Audit.

[69] *Julia Lawrence v AG Grenada* [2007] UKPC 18.

[70] (2004) HC 18. Unreported. Available on file with the author.

7.3.2 Small Size and Political Patronage

As our detailed analysis of the constitutional provisions surrounding the appointment of the key constitutional actors has shown, the Prime Minister's involvement in the process is ubiquitous. Of course, it might be argued that this is not unique to the Commonwealth Caribbean: there are many other examples of Commonwealth countries where the Prime Minister is vested with very extensive power of appointment and dismissal. However, the problem of prime ministerial patronage assumes a particular shape and form in the region because of the small size of the majority of its counties and their faltering economies. In these small and relatively undeveloped countries the dynamics of the relations between the Prime Minister and the other main constitutional actors are much more unequal when compared with those in larger more developed countries elsewhere in the Commonwealth. Even if the dimensions of government are not significantly greater, the state occupies a disproportionate space within the life of its citizens in small countries.[71] The opportunities for career advancement outside the public sector are very limited, and the end of a career in public office may often coincide with the end of earning a livelihood. There are, of course, always the occasional mavericks, such as the Auditor General of Grenada, discussed above, and the Chairman of the Electoral Commission of Antigua and Barbuda, discussed below, who are prepared to stand up to the Prime Minister, but they are very much the exception to the general rule. The overwhelming tendency, unsurprisingly, has been for public officials not to be too critical of the person who was responsible for placing the individual in a position of privilege in the first place.

Prime Ministers in the region are thus able to extract an unusually high level of loyalty and deference from those whom they have appointed, which conflicts with the latter's willingness and ability to discharge their constitutional duties. As we will see below, in such circumstances the Courts have, on a number of occasions, been asked to intervene either to oblige these officials to discharge their duties in accordance with the constitution, or to prevent Prime Ministers from transgressing the limits imposed on them by the constitution

7.4 The Role of the Courts

There is not the space here to review each and every case in which the Courts have been asked to intervene to prevent abuse of power by Prime Ministers or public officials,[72] so I propose to focus instead on four recent examples, drawn from St Kitts and Antigua and Barbuda, which illustrate the problem of prime ministerial influence in the context of the conduct of elections, and in three of which the Courts

---

[71]Transparency International (2004), p. 20.
[72]For further examples see O'Brien (2014).

were required to intervene. Elections are often cited as a vital sign of the region's commitment to democracy, and successive governments have, since independence, respected the outcome of elections by peacefully surrendering power to their successors. However, the 'winner takes all nature' of the first-past-the-post electoral system, means that the stakes for Prime Ministers are particularly high as is the temptation to exert their residual influence over election officials.

Both of the countries that I have chosen share enough constitutional features in common with other countries in the region to suggest that they are reasonably representative, and the examples that I have selected provide graphic case studies of the kind of abuse of power that can result from the type of relationship between Prime Ministers and their key officials that is all too common in the region.

7.4.1 Reviewing Constituency Boundaries

The recent case of *Brantley and Others v CBC and Others*,[73] on appeal from St Kitts and Nevis, illustrates perfectly the tension that exists between the CBC's duty to periodically review constituency boundaries and its relationship with a Prime Minister who recommends the appointment of both the Chairman and two of the other four members of the CBC.[74]

On this occasion, the Opposition, upon learning that the Government was proposing to introduce certain boundary changes recommended by the CBC prior to elections that were due to be held in 2015, notified the Prime Minister of their intention to challenge the recommendations on the ground that they did not comply with the requirements of Schedule 2 of the Constitution.[75] Thus alerted to the Opposition's intentions, the Government responded extraordinarily quickly. Within the space of 5 hours, beginning at 2 pm on 16th January, the Government managed: to have the draft report of the CBC signed off by a majority of its members (the Opposition members of the CBC having refused to sign the draft report); to hold an emergency meeting of the National Assembly to approve a draft proclamation giving effect to the CBC's report; to have the draft proclamation then signed by the Governor General at the same time as the Governor General signed a proclamation dissolving Parliament; and, finally, at about 6.30 pm, to screen a broadcast by the Prime Minister, announcing that the impugned proclamation had been 'gazetted',[76] that Parliament had been dissolved, and that a general election would be held on 16th February 2015.

---

[73][2015] UKPC 21.

[74]Constitution of St Kitts and Nevis, s 49(1).

[75]Schedule 2 provides, inter alia, that there should be an equal number of inhabitants in each constituency as far as is reasonably practicable always having regard, inter alia, to the need to ensure adequate representation of sparsely populated rural areas.

[76]In accordance with the Constitution of St Kitts and Nevis, s 119.

The speed with which the Government acted was a function of the ouster clause contained in section 50(7) of the Constitution. This provides that once the report of the CBC has been approved by the National Assembly, and a proclamation has been 'made' by the Governor General, the validity of the proclamation cannot be enquired into in any Court of law. The Government thus hoped that if they acted quickly enough the Opposition would be prevented by the ouster clause from mounting a legal challenge to the boundary changes.

The Government's efforts to rush through the boundary changes were ultimately, however, to no avail. This was because upon an appeal by members of the Opposition to the Judicial Committee of the Privy Council (the 'JCPC')[77] it was held that the impugned proclamation by the Governor General had not been 'made' in accordance with section 50(6) of the Constitution until it had been published in the Gazette (in the sense of a hard copy of the Gazette being available to the public); broadcasting the proclamation on the radio and television being insufficient to satisfy section 50(6). On the basis of the unchallenged evidence adduced by the appellants, the proclamation was not 'made' before the 20th January 2015 at the earliest. Since section 50(6) also provides that the draft proclamation by the Governor General only comes into effect upon the next dissolution after it has been 'made', it followed that the dissolution of Parliament, which took effect from the 16th January 2015, predated the 'making' of the impugned proclamation. As a result the impugned proclamation, even if valid, would only take effect on the dissolution of the Parliament that was elected on 16th February 2015. This meant that the election which was due to be held on 16th February had to be fought on the basis of the boundaries existing before the purported alteration on 16th January 2015.

7.4.2   The Announcement of Election Results

The Government's failure to implement the changes recommended by the CBC was not, however, the end of the matter. The subsequent elections, in which the Opposition won 7 out of the 11 available seats, were described by one eminent Caribbean commentator, Sir Ronald Sanders, as 'a fiasco', due to the failure of the Supervisor of Elections to declare the results until 2 days afterwards. As Sir Ronald Sanders noted, there were only 30,000 voters in the election and even if the votes had been counted twice for accuracy, as the Supervisor of Elections had claimed, a final count should have been available by midnight on the day of the election at the latest. Though it was never conclusively proven, there was a widespread suspicion, based on the fact he had been appointed on the recommendation of the Prime Minister, that the Supervisor of Elections had delayed the announcement of the

---

[77]The JCPC remains the final appellate Court for the majority of countries in the region. The exceptions are Barbados, Guyana, Belize and, most recently, Dominica.

election results at the Prime Minister's behest. The former Prime Minister of St Kitts and Nevis, Kennedy Simmonds, for example, was stinging in his criticism of the delay: 'What we are seeing here...is totally unprecedented...we are once again being made the laughing stock all over the Caribbean and the world'.[78] The Prime Minister of Trinidad and Tobago, Kamla Persad-Bissessar, was also concerned about the implications of the delay for the region's reputation for democracy and for free and fair elections:

> As a region we have to be very careful of the messages that we send and that which is emanating from St Kitts and Nevis is not the kind of message we want to send to our people and to the world.[79]

7.4.3   Reconfiguring the Composition of Electoral Commissions

We have seen above how the Governor General must consult with the Prime Minister when appointing the Chairman of the Electoral Commission. Prime Ministers may also be involved at the preliminary stages in the removal of the Chairman when recommending the establishment of a tribunal to enquire into the question of whether the Chairman of the Electoral Commission should be removed. Sometimes, however, a Prime Minister may seek to have an even more direct input into the Chairman's removal, as occurred in Antigua and Barbuda, following the 2009 elections.

In this case the Antigua and Barbuda Electoral Commission (ABEC) had been subject to a barrage of criticism from the governing United Progressive Party (UPP) in the run up to the 2009 elections. Even though the UPP was ultimately victorious in the elections, the Prime Minister remained unhappy with the handling of the elections by ABEC and its Chairman because of the late opening of the polls in a number of constituencies,[80] which the Prime Minister believed had cost his party a number of seats. Following the election, the Prime Minister recommended that a tribunal be established by the Governor General to consider whether the Chairman should be removed. Though the tribunal subsequently issued a report commending the Chairman and advising that it did not recommend his removal, the Prime Minister was determined to have his way and instructed the Governor General summarily to remove the Chairman. Even though this was clearly in breach of the Constitution, the Governor General, nevertheless, acted upon the Prime Minister's instructions and removed the Chairman from his post.[81] In response the Chairman filed proceedings for judicial review and, though he was unsuccessful at first

---

[78]Douglas (2015).

[79]Ibid.

[80]See Statement of ABEC available at http://www.abec.gov.ag/pr/ER_2009.pdf

[81]In accordance with the Representation of the People (Amendment) Act 2001, s 4.

instance, his appeal was ultimately upheld by the Eastern Caribbean Supreme Court ('ECSC'),[82] which had no hesitation in holding that the decision of the Prime Minister to recommend his removal as Chairman was illegal, irrational and procedurally unfair; and that the decision of the Governor General to remove the Chairman was also ultra vires and, therefore, unlawful.

The Prime Minister was not, however, deterred by this judgement. Anticipating that he might lose the appeal before the ECSC, the Prime Minister had already taken the precaution of enacting the Representation of the People (Amendment) Act 2011 before the ECSC had even delivered its judgement. Section 1 of the Amendment Act 2011 provided that it should come into force on such day as the Prime Minister might appoint by Order. On 31st January 2012, 6 days after the ECSC delivered its judgement, declaring the recommendation of the Prime Minister to remove the Chairman as unlawful, the Prime Minister appointed 22nd December 2011 as the day on which the Amendment Act 2011 would be deemed to have come into force. The effect of the Prime Minister's Order was to nullify, at a stroke, the judgement of the ECSC: dissolving the existing Commission and empowering the Prime Minister to recommend the appointment of a new Chairman. Once again, however, the deposed Chairman responded by commencing proceedings seeking judicial review of the Prime Minister's Order. Though unsuccessful at first instance, the Chairman's appeal was eventually upheld by the ECSC, which adjudged that the Prime Minister had acted unlawfully in selecting the date that he chose for the Order to come into effect, with the clear intention of undermining or undercutting the ECSC's earlier judgement in favour of the Chairman. The Order was, accordingly, declared to be null and void and of no legal effect.[83]

In a final twist to this saga, following the UPP's loss in the subsequent 2014 general election, and with the ABLP now forming the majority in the House of Representatives, the former Chairman of ABEC, who had proven to be such a thorn in the flesh of the UPP and the deposed Prime Minister, was elected to be Speaker of the House of Representatives, at a meeting of the House which was boycotted by the UPP.

7.4.4 No-Confidence Votes

Under the Westminster model the executive is formed by and is, ultimately, accountable to the legislature through the convention of collective ministerial responsibility. This means that a government which loses a vote of no-confidence

---

[82]*Watt v Attorney General and Prime Minister Antigua and Barbuda* No. ANUHCV 2011/0025. Unreported decision of the Eastern Caribbean Supreme Court. Available on file with the author.

[83]*Watt v Prime Minister Antigua and Barbuda* No. ANUHCVAP2012/0042. Unreported decision of the Eastern Caribbean Supreme Court. Available on file with the author.

in Parliament is expected immediately to resign. Votes of no-confidence in the region are, however, rare for at least two reasons.

The first is the power of the Prime Minister to request a prorogation of Parliament. While there has been much academic debate about the circumstances in which a Governor General might refuse a Prime Minister's request for a prorogation, even those who argue that the Governor General has no power to refuse such a request do so on the assumption that any such prorogation will be a temporary affair, a matter of weeks, and that the Opposition will always have an opportunity to table a motion of no-confidence at the next sitting of Parliament.[84] However, in the Commonwealth Caribbean there has been at least one example of a Prime Minister requesting an indefinite prorogation of Parliament. This occurred in Grenada, in 1989, when the Prime Minister, Herbert Blaize, acted to pre-empt the possibility of a defeat on a vote of no-confidence by requesting the Governor General to prorogue Parliament indefinitely until such time as the Prime Minister was ready to request the Governor General to dissolve Parliament and fix a date for a new election. Though it is arguable that one example does not set a precedent, the concern is that the nature of the relationship between the Governor General and the Prime Minister in these small countries ensures that the Prime Minister's wishes will always prevail, even if the prorogation is for an indefinite period, and even if there are legitimate doubts about the Prime Minister's motives for requesting a prorogation.

The second reason why votes of no-confidence are rare is the ability of the Government to control the parliamentary schedule through its influence over the Speaker, who is elected by the majority party in the legislature, and who may refuse to afford time to debate a motion of no-confidence for as long as possible, as occurred recently in St Kitts and Nevis. In this case a majority of the members of the House of Assembly, entitled to vote, first submitted a motion of no-confidence in December 2012.[85] Despite an assurance by the Speaker that the motion would be heard as soon as possible, and despite numerous requests by members of the Opposition, the motion had still not been debated by April 2013. Faced with the refusal of the Speaker to table their motion, members of the Opposition filed an originating motion in the High Court seeking, inter alia, an injunction requiring the Defendants—who included the Prime Minister, his Cabinet and the Speaker of the National Assembly—to take whatever steps were necessary to ensure that the motion of no-confidence was debated as soon as may be practicable.

The outcome of these proceedings was, however, something of a pyrrhic victory for the Opposition. While the Court accepted that it had jurisdiction to review the actions of the National Assembly to ensure that they were consistent with the Constitution, and that each member of the Assembly has a right to request that a motion of no-confidence be debated and voted on within a reasonable time as a matter of priority, it also held that there was nothing in the pleadings to ground an allegation that the Prime Minister or members his Cabinet had prevented the

---

[84]See MacDonald and Bowden (2011).

[85]Robinson et al. (2015), p. 103.

Speaker from tabling the motion of no confidence for debate.[86] On this basis, the proceedings against the Prime Minister and his Cabinet were struck out, though the claimants were allowed to proceed with their substantive motion against the Speaker.

With the benefit of hindsight it is arguable that the decision by members of the Opposition to commence proceedings against the Prime Minister and the Speaker to force a vote of no-confidence was politically naive. The Prime Minister of St Vincent and the Grenadines, Ralph Gonsalves, has, for example, observed, that the Opposition played straight into the Government's hands in commencing legal proceedings because it allowed the Prime Minister to sit back and await the outcome of the substantive motion against the Speaker, knowing that there was absolutely no prospect of either this motion or the motion of no-confidence being heard prior to the 2015 elections. It is, however, difficult to know what other remedy was open to the Opposition in a National Assembly that was dominated by members of the Government and presided over by a Speaker who had been elected by members of the governing party. As one senior lawyer observed at the time, the whole handling of the no-confidence vote, 'strikes at the core of parliamentary democracy and does no credit to the Speaker of our National Assembly and the present Government'.[87]

7.5   Conclusion

The post-independence experience of the Commonwealth Caribbean calls into question the rival claims both of those scholars who claim that democracy is a function of a country's small size and those scholars who claim that it is a function of a country's historical circumstances, especially where that country has enjoyed the benefits of British colonial rule. In the case of the Commonwealth Caribbean it would be more accurate to say that British colonial rule and the small size of the majority of its countries, when combined with the enormous power vested in the Prime Minister, have together produced a political culture which has more in common with the autocratic rule of a colonial Governor than it does with a system of democratic, responsible and accountable government.

As we have seen, the contingent status of Governors General in combination with their close political and social ties to the Prime Minister can make it difficult for them to exercise their powers independently, even when called upon by the constitution 'to act in their own deliberate judgment'. Prime Ministers are also firmly in control of the legislature. Lower Houses in the region, which are usually small in number, are frequently dominated by members of the Cabinet who are bound by the principle of collective responsibility to support the Prime Minister.

---

[86]*Brantley v Martin* (2014) HC SKN. Unreported. Available on file with the author.

[87]Ferdinand (2013).

Proceedings in Lower Houses in the region are also regulated by a Speaker appointed by the party forming the majority, and at least one Speaker has been known to refuse to table for debate a motion of no confidence which, had it been tabled, would have forced the Prime Minister to resign. Upper Houses, meanwhile, are typically composed of a majority of members appointed upon the recommendation of the Prime Minister and can be removed upon the recommendation of the Prime Minister for not supporting the government's legislative programme. In addition, Prime Ministers across the region are involved in the appointment of members of the PSC, members of the Election Commission and CBC, and the Auditor General; in all three cases undermining the political neutrality of these key public officials.

This lack of political accountability means that the Courts have often been required to intervene to hold both the Prime Minister and other senior public officials legally to account. We have seen, for example, how members of the Opposition were obliged in the case of St Kitts and Nevis to ask the Courts to overturn a decision of the CBC, which had been railroaded through by the Government immediately prior to the 2015 elections. We have also seen how the Chairman of the Election Commission in Antigua and Barbuda turned to the Courts for protection after the Prime Minister instructed a compliant Governor General to ignore the recommendation of a tribunal that the Chairman should not be removed; and, when this failed, introduced ad hominem legislation to secure the Chairman's removal.

In these cases, the Courts undoubtedly played an important role in upholding democracy and the rule of law. However, the fact that the Courts were involved in the first place challenges us to question traditional assumptions about the ways democracy functions in small countries. It also challenges us to question assumptions about the ways in which centuries of British colonial rule have shaped the political culture of these former colonies in the postcolonial era.

References

Anckar D, Anckar C (1995) Size, insularity and democracy. Scand Polit Stud 18:211–229

Antigua Constitutional Review Commission (2002) Report. Antigua

Barrow-Giles C, Joseph TSD (2006) General elections and voting in the english-speaking Caribbean 1992–2005. Ian Randle Publishers, Kingston

Blackstone W (1765) Commentaries on the laws of England, vol 1. Clarendon Press, Oxford

Bogdanor V, Marshall G (1996) Dismissing governor-generals. Public Law:205–213

Dale W (1983) The modern commonwealth. Butterworths, London

Diamond LJ, Tsalik S (1999) Size and democracy: the case for decentralization. In: Diamond LJ (ed) Developing democracy: towards consolidation. John Hopkins University Press, Baltimore

Dominguez JI (1993) The Caribbean question: why has liberal democracy (surprisingly) flourished? In: Dominguez et al (eds) Democracy in the Caribbean. John Hopkins University Press, Baltimore

Douglas S (2015) Dr Douglas Voted Out. Trinidad and Tobago Newsday. Available at http://www.newsday.co.tt/news/0,207085.html. Accessed 01 July 2016

Dupont J (2001) The common law abroad: constitutional and legal legacy of the British Empire. Fred B Rothman Publications, Littleton

Eaton G, et al (2002) The Public Service Commission in Grenada. In: Ryan S, Bissessar A-M (eds) Governance in the Caribbean. SALISES - University of the West Indies Press, St Augustine

Faris R (1999) Unto themselves: insularity and democracy. Thesis (MA – Unpublished), University of North Carolina

Ferdinand E (2013) Commentary by Emile Ferdinand QC: constitutional conventions and political controversies. Winn 98.9 FM

Ghany H (2002) Defeated parliamentary candidates as parliamentarians: a comparative study of political will vs electoral will in Grenada and Trinidad and Tobago. Paper presented at the Grenada conference - beyond walls: multi-disciplinary perspectives. University Centre, St. George's, 7–9 January 2002

Holland Rose J (ed) (1940) Cambridge history of the British empire, vol 2: the growth of the new empire 1783–1870. Cambridge University Press, London

Kumarasingham H (2010) Onward with executive power: lessons from New Zealand 1947–1957. Wellington, Institute of Policy Studies, Victoria University of Wellington

Lewis GK (2004) The growth of the modern West Indies. Ian Randle Publishers, Kingston

MacDonald NA, Bowden JWJ (2011) No discretion: on prorogation and the Governor General. Can Parliam Rev 4(1):7–16

Meighoo K, Jamadar P (2008) Democracy and constitution reform in Trinidad and Tobago. Ian Randle Publishers, Jamaica

O'Brien D (2014) The constitutional systems of the commonwealth Caribbean: a contextual analysis. Hart Publishing Ltd, Oxford

Ott D (2000) Small is democratic: an examination of state size and democratic development. Garland, New York

Payne DW (1993) Caribbean democracy. Free Rev 10:1–19

Robinson T et al (2015) Fundamentals of Caribbean constitutional law. Sweet & Maxwell, London

Srebnik H (2004) Small island nations and democratic values. World Dev 32:329–341

Transparency International (2004) Country Study Report: Caribbean Composite Study. Berlin

Veenendaal WP (2013) Democracy in microstates: why smallness does not produce a democratic political system. Democratization 22(1):92–112

Wrong H (1923) Government of the West Indies. Clarendon Press, Oxford

Chapter 8
Legal Pluralism and Politics in Samoa: The Faamatai, Monotaga and the Samoa Electoral Act 1963

Tamasailau Suaalii-Sauni

8.1 Introduction

On the 3rd of April 2016, an article was published by Samoa's popular national newspaper, the *Samoan Observer*, titled: 'Associate Minister wanted'. The short text, under the large photo of the wanted former Minister, read:

> The former Associate Minister of the Prime Minister's Office, Lafaitele Patrick Leiataualesa, is a wanted man. Prime Minister, Tuilaepa Sa'ilele Malielegaoi has been searching for him.
>
> Speaking to the media, Tuilaepa said he has been trying to find his Associate Minister in relation to a petition he filed against another Member of Parliament of the Human Rights Protection Party (H.R.P.P).
>
> Lafaitele is taking newly elected woman M.P for Alataua West, Aliimalemanu Alofa Tuu'au to Court over allegations of bribery and treating.
>
> It is one of six petitions filed with the Ministry of Justice and Courts Administration.
>
> But the leader of the H.R.P.P. is not happy with this.
>
> "It's not a problem if it's ([a] petition) from another party", said Tuilaepa.
>
> "I have been looking for my Associate Minister who filed [the] petition and (I was told) he's in America but I searched the internet and I found that he is still in Siusega".
>
> ... Tuilaepa added that it's shocking that some of those that have filed petitions hold positions in church.
>
> ... "It's an internal matter", he explained. "I believe that time will pass and the candidate will be reminded that it is between them and a brother, and might reconsider to put it aside...especially being from the same party and not between us and the Opposition party that used to be".
>
> He [the PM] reminded that there is still time and believes that God is speaking to some of the petitioners – including his former colleague Lafaitele.[1]

---

[1]Tupufia (2016a).

T. Suaalii-Sauni (✉)
Sociology/Criminology, School of Social Sciences, University of Auckland, Private Bag 92109, 10 Symonds Street, Auckland, New Zealand
e-mail: s.suaalii-sauni@auckland.ac.nz

© Springer International Publishing AG 2017
P. Butler, C. Morris (eds.), *Small States in a Legal World*,
The World of Small States 1, DOI 10.1007/978-3-319-39366-7_8

I am told by friends and family in Samoa that the search for the missing former Associate Minister is not only being conducted quite publicly by the Prime Minister, but also by the former Associate Minister's village fono.[2] In fact, this former Associate Minister was suspended from his H.R.P.P party for 12 months in 2013, allegedly for swearing at the organisers of a formal ava ceremony for getting (in the opinion of the former Associate Minister) the distribution sequence of ava cups wrong. One of his fellow H.R.P.P members went to bat for him in an H.R.P.P. caucus meeting saying that: 'In Samoan custom if there is an error in the distribution of ava or in what is said by a matai involved, the offenders would be subjected to the strongest swear words from the elders amongst the matai'.[3] The suggestion was that by Samoan custom Lafaitele was not wrong for swearing and should not have been suspended. Nevertheless, Lafaitele's suspension held.[4]

Apart from the usual political shenanigans this newspaper story introduces, it also highlights three things of particular interest for our purposes. Firstly, is the importance of being able to read cultural nuance in these socio-political reports and events and its relevance to understanding custom. Samoa's recent H.R.P.P landslide election victory means that Samoa effectively has a one party state where law-making will be dominated by H.R.P.P persuasions over the next 5 years.[5] Without the checks and balances offered by an opposition party, Samoa's voting public must find other ways to hold Parliament and the government accountable to its prized rule of law—a rule of law assumed to be capable of reading, knowing and giving due regard to the nuances of custom and culture, i.e. to Samoa's faasamoa, aganuu, agaifanua, and tu ma aga,[6] as hoped for by the Constitution.

My second point for attention is the potential negative effects caused by the ambiguities created by the ad hoc blending of Samoa's faamatai (chiefly) and parliamentary democratic systems. In Samoa all parliamentarians must by state law hold a matai or chiefly title.[7] This puts the faamatai squarely within the decision-making whirlpool of Samoa's Westminster parliamentary style

---

[2]Fono is a Samoan term for a meeting or council. Village fono refers to a village council. In Samoa the Village Fono Act 1990 was enacted to give state recognition of the importance of the village fono to local governance, among other things.

[3]Keresoma (2013).

[4]Ibid.

[5]Official election results released by Radio New Zealand on 12 March 2016. H.R.P.P won 47 of the 50 seats, demolishing the former Tautua opposition party, leaving them with only 3 members in the current Parliament. See Radio New Zealand (2016a).

[6]The terms faasamoa, aganuu, agaifanua and tu ma aga, are variously used to refer to Samoan customs, culture, traditions and worldview. Faasamoa, literally means 'to be of Samoa'; aganuu refers to the ways of the village or community; agaifanua refers to the ways of the land of one's birth—land in this case is most commonly understood to refer to the place of birth or birth village, but may also refer to the place to which the chiefly title belongs; tu ma aga refers to customs, literally 'tu' means the place in which one's feet stands, i.e. 'tu' is to stand, from the phrase 'tulaga vae'—vae refers to feet; tulaga refers to position or positioning of, 'aga' refers generally to the idea of 'the ways of a people or group'. See Tamasese (2016).

[7]Electoral Amendment Act 1990.

democracy. Knowing, among other things, how to navigate the codes of conduct required of a matai as opposed to a parliamentarian or both requires deliberate examination of what these codes are in theory and in practice. This opens the gates to an analysis of the historical and ideological foundations of both systems—where they meet and where they do not. Much is known about the philosophical bases of the jurisprudential traditions of the common law and Westminster politics.[8] Very little deliberate scholarly examination is, however, available on an indigenous jurisprudence for Samoa.[9]

My third point for attention is the continuing presence and dominance of God— of both God Ieova[10] and God Tagaloa[11]—in Samoa's contemporary governing discourses and the lack of attention that theology has received in examinations of legal pluralism in the Pacific.[12]

All three of these points, i.e. the reading of cultural nuance in law, the blending of the faamatai and parliamentary democracy, and the co-existence of God Ieova and God Tagaloa in Samoan politics and law—raise interesting claims about Samoa's legal pluralism, its texture and character, and about Samoan party politics and how it informs and reflects that texture and character. This chapter therefore seeks to discuss the claim that Samoa has a dual legal system. It explores how this claim is shaped by the way in which legal pluralism is currently defined and how an understanding of it in Samoa is informed by the interplay between the faamatai (Samoa's chiefly system), Samoa's parliamentary system, the faasamoa (Samoa's customary system), and the faakerisiano (Samoa's Christian system).

My overall argument is that Samoa, like many other Pacific island states, has, for various understandable reasons, neglected the development of its custom law body of knowledge. Because of this there is no properly established corpus of knowledge

---

[8]Simmonds (2008), Harris (2002), and McLeod (2007).

[9]The recent works of His Highness, Samoa's current head of state, Tui Atua Tupua Tamasese Ta'isi Tupuola Tufuga Efi are seminal in this regard, largely for the depth of analysis he gives to custom. As a cultural custodian Tui Atua was schooled in the customary knowledge of many of Samoa's leading orators. Much of this knowledge was tapu (or sacred and protected) knowledge. As Tui Atua writes, in publishing some of the tapu knowledge passed on to him by his mentors he is breaking tapu and decided to do so notwithstanding in order to save much of what would otherwise be lost. See Suaalii-Sauni et al. (2009). The work of the late Asiata Saleimoa Vaai is also of considerable note. See Vaai (1999). Asiata was a trained lawyer and a political leader of Samoa. He was leader of the Samoa United Independents Political Party and the Samoa Democratic United Party.

[10]Ieova is a Samoan loan translation of the term Jehovah, which is in turn the English loan translation of the Jewish name for their God, Yahweh.

[11]God Tagaloa is in reference to the anthropomorphic God of Samoa that is also commonly referenced as the anthropomorphic God of other Polynesian countries as recorded in Polynesian oral histories. See Meleisea et al. (1987) and Marck (1996).

[12]The recently published edited collection, titled Whispers and Vanities: Samoan indigenous knowledge and religion, offers useful discussion on some of the theological tensions between faakerisiano (Christian beliefs) and faasamoa (Samoan beliefs) about God in Samoa and other parts of the Pacific, particularly within Polynesia. See Suaalii-Sauni et al. (2014).

available to scholars to make an informed assessment of whether, in juristic terms, it is a legally plural state or not.[13] While it may be premature to declare Samoa a country with a dual legal system, there is ground to argue that there exist three sources of law both recognised by the people of the modern-day independent state of Samoa and by their Constitution,[14] all of which draw legitimacy from different normative orders. This suggests that while on an empirical social science basis a state of legal pluralism can be said to exist (i.e. there are plural sources of law that do not need the state for legitimacy and are behaviourally recognised by the people in their everyday lives), juristically speaking the issue is complicated. Achieving juristic legal pluralism was definitely an ideal dreamt of by the drafters of the Constitution and it is a dream not yet discarded by Samoans, the author included. However, to realise this dream a sophisticated body of scholarship must be developed with the sole purpose of unpacking the juristic elements of the faasamoa and making it speak to wider jurisprudential concerns. This article hopes to contribute to that.

To begin with I contend that an exploration of party politics in Samoa and the recent codification of the Samoan custom of monotaga can lend insight into how this unpacking exercise might occur and at the same time offer the kind of 'circumstantial specificity' and 'ideological texture' inherent in legal pluralism that Sally Engle Merry argues must be brought out in order for us to better understand the character of law and legal pluralism.[15] Before delving into this, it is instructive to offer a brief outline of the concept of legal pluralism as it applies here.

8.2   Legal Pluralism in Samoa

The New Zealand Law Commission in their analyses of legal pluralism and its significance to understanding custom and human rights in the Pacific, noted that for legal pluralists 'law derives from people, and state-made law gives expression only to one form of it' and that custom is presented as law 'and as constituting a legal

---

[13]This neglect was contributed to by a number of structural and attitudinal factors: 1. The difficulties associated with having access to tapu knowledge (which a lot of customary knowledge is), mostly because cultural custodians either took the knowledge with them to their graves or they would refuse to part with it in any coherent way; 2. A lot of this knowledge was also lost when a significant proportion of the Samoan adult population died during the influenza epidemic that swept Samoa in 1918 during the New Zealand administration of Samoa; 3. With missionary influence a lot of this knowledge was discarded in favour of Christian-oriented beliefs when Samoans converted to Christianity; and 4. A misplaced presumption that Samoan indigenous knowledge could never be lost as long as the Samoan language survived and Samoa was independent of foreign rule.

[14]See the Constitution of the Independent State of Samoa 1960. There are Samoan and English language versions of the Constitution.

[15]Merry (1988).

system that is not dependent on state recognition for its authority'.[16] On this definition Samoa's legal situation would seem to reflect one of legal pluralism as so defined. Teleiai Dr. Lalotoa Mulitalo's doctoral thesis, which explores the consequences of legal pluralism on law reform in the South Pacific, agrees with this profile.[17] Although she does accept that '[a] central challenge...is the lack of systems in place to accommodate the development of a law reform framework suitable for legal pluralism'.[18] And, that customary law, its juristic elements, are paid 'scant regard' in current law reform systems.[19]

Almost 20 years ago Merry (1988) examined how legal pluralism, as a concept, has both social and legal dimensions and that while there are analytical and systemic advantages to differentiating between these for juristic purposes on the one hand, and social science on the other, it is generally accepted by scholars from both camps that legal pluralism offers a useful point of focus for examining the complex interplay between law and society. Knowledge of the relationship between law and society has evolved such that the idea of having more than one legal system (in the sense offered by Pospisil)[20] operating alongside or within each other is more or less accepted as part of the continuum of views on what is legal pluralism. The legal pluralism of interest here is that which identifies custom law (and the ten commandments for that matter) as legitimate sources of law, whether or not it is accepted into state law by codification into an Act of Parliament. This leaves aside for the moment the argument that law is ultimately about the power of one source to make another subservient to it.[21] Legitimacy in and of itself lies either with the people or with Parliament or both. But how do we know how 'the people' understand law and its role in contemporary Samoa?

In the context of Samoa, Samoa's Constitution is generally accepted by the nation as the supreme law of the land. And, like other small Pacific island states, the Constitution recognises not only Samoan custom and tradition (tu ma aganuu a Samoa), but also Christian principles as (potential) sources of law.[22] Exploring the question of the Bible as a source of law falls outside the purview of this chapter. The question of theology inevitably arises, however, if one delves deep enough into the cultural, philosophical, epistemological and ideological premises of Samoan

---

[16]New Zealand Law Commission (2006), p. 42.

[17]Mulitalo (2013) and Mulitalo (2012a), p. 28: where she discusses the challenges of drafting legislation in Samoa, a country where she argues legal pluralism prevails. Mulitalo has worked as a legislative drafter for the Samoan government.

[18]Ibid.

[19]Ibid.

[20]Merry (1988), p. 870.

[21]In ways both coercive and productive as suggested by Foucault in his theses on discipline and punishment, and governmentality. See Foucault (1979) and Foucault (1994), pp. 229–245.

[22]These Samoan words are used in the preamble of the Constitution. See discussion of this in Suaalii-Sauni (2010), pp. 70–88.

custom, and the traditional and contemporary practice of these.[23] I make mention of it here merely to underline the point about Samoa's plural sources of law and what gives it richness, texture and flavour. Probing the theological premises of law in Samoa has for various reasons been avoided by modern Samoan legal scholars.[24] As a critical part of Samoan customary law and of contemporary practices, the theological question ought to be properly explored in Samoan law reform frameworks and law and society research.

Jennifer Corrin's call for a more intuitive or nuanced understanding of the 'complex interplay between the interwoven spheres of "traditional law" and "state law"', and of what she describes as 'a new sphere of "blended law"', is of value to note in terms of working through what legal pluralism is or ought to be.[25] This 'blended pluralism' emerges or is realised, she implies, once a view of legal pluralism is opened up to allow for constructions of law beyond those employed by dichotomous or hierarchically oriented models. As Corrin correctly points out, while custom law and state law do co-exist in Pacific island states, they do so in ways that are arguably both blended and parallel. This refines our way of thinking about law and legal pluralism as a legal and social concept. It presents legal pluralism as a phenomenon that currently includes three different definitions and/or positions within it, as described in Table 8.1.

The question, of whether Samoa is a legally plural state or not matters because if the question of legal pluralism can be properly scrutinised it could provide the very foundation upon which Samoa can develop the blueprint advocated for by Kenneth

Table 8.1 Legal pluralism

All law sources recognised by people are legitimate and co-exist in parallel spheres	All law sources recognised by people are legitimate and operate in both parallel and blended ways	Law is hierarchical; lower tier law sources must have approval of top tier

---

[23]The history of the separation of Church and State in the history of western legal liberalism is not a history shared by Samoa and many of the other Pacific island nation states. For Samoa at the time of independence the adoption of Westminster models of justice and governance was more a matter of political necessity than widespread ideological belief. This creates significant theoretical and practical tensions for how the very notions of democracy and the rule of law ought to operate and why. Without specific address of the theological underpinnings of all that informs a Samoan jurisprudence, whether considered plural or otherwise, scholars and practitioners would miss this very crucial point.

[24]In fact there are only a handful of notable Samoan legal scholars that conduct deliberate analyses of Samoan custom as a legitimate source of law. On top of the work of Teleiai Lalotoa Mulitalo, Asiata Saleimoa Vaai, and His Highness Tui Atua, is the excellent work of Fanaafi Aiono-Le Tagaloa. See Aiono-Le Tagalog (2009). The work of Samoan theologians such as Lalomilo Kamu on faasamoa and the Christian gospel has also been drawn on by Samoan social scientists interested in the philosophical premises of Samoan customary concepts, principles and institutions. See mention of Lalomilo Kamu's work in Huffer and So'o (2005), p. 313.

[25]Corrin (2009), p. 29.

Brown (1999). It forces Samoa to consider what societal norms and values she wants protected for her children and for her children's children, why and how. There is little doubt that Samoa wants her customs and traditions protected, that her Christian beliefs are also of significant value and that being able to actively participate in the global economy and meaningfully contribute to the development of humanity is important to her. Ensuring that all this is protected in law, however, requires taking time to spell out the nuances of what customs and traditions mean and ought to mean in the Samoan contemporary legal context. The custom law of monotaga and its codification in Samoa's state law (the Electoral Act) is offered as an illustration of both the complex nature of legal pluralism as a concept and of how custom is treated in contemporary Samoan legal practice.

A useful aspect of legal pluralism is its ability to directly challenge both the restrictiveness of legal positivism and the monoculturalism of Westminster-oriented common law.[26] In the Pacific this is most pronounced in countries where introduced laws are still used without adaptation to the local context, and in cases where if there is a conflict of meaning the foreign language version is ruled by judges to be preferred over the indigenous language version, as discussed below.[27] The languaging of the English version of the new Village Fono Amendment Bill 2015, suggests that change is in the air.[28] However, current legislative definitions of the custom of monotaga and judicial interpretations of the same suggests that this change is both slow and selective.

There has been much written on the faamatai, on Samoan politics, Samoan elections, and Samoan law, including customary law. However, very little of this has focused on how they come together to help us understand the nature of legal pluralism in Samoa. This is my focus. I turn now to first offer some background information on the small island state of Samoa before discussing the faamatai and Samoa party politics.

8.3 The Independent State of Samoa, the Faamatai and Party Politics

Samoa's political and legal sovereignty extends to two large islands, Upolu (which is approximately 1114 km^2) and Savaii (which is approximately 1820 km^2). There is included also the much smaller islands of Apolima, Manono, Fanuatapu, Namua,

---

[26] Corrin (2016a).

[27] Angelo (2012). See also Article 112 of the Samoa Constitution.

[28] The English version of the Village Fono Amendment Bill 2015, s 2, which seeks to amend the Village Fono Act 1990, defines the legal terms 'faiga fa'avae' or 'i'ugafono' to mean 'village faiga fa'avae or i'ugafono made pursuant to section 5'. The decision to retain the indigenous language in the defining of the indigenous terms rather than to offer an English translation is a major step forward in claiming legal pluralism in a juristic sense in Samoa.

Nu'utele, Nu'ulua and Nu'usafe'e. Of these smaller islands only Apolima and Manono are inhabited. The Aleipata islands (Namua, Nu'utele, Nu'ulua and Fanuatapu) have been identified by the Samoa Ministry of Natural Resources and Environment (MNRE) as an important habitat for seabirds, the ground dove and the hawksbill turtle, and so these are protected by law.[29]

According to Samoa's Statistics Bureau there are approximately 195,000 people living in Samoa today.[30] The independent state of Samoa is proud to be the first island nation in the Pacific to rid itself of its colonial imperial past. In 1962 it gained full independence from New Zealand and was recognised as a nation state, and one quite separate—at least legally—from Eastern Samoa (more commonly known as American Samoa[31]). In 1997 Samoa's then Parliament decided to also rid Samoa of the preface 'Western', preferring that Samoa be named just Samoa.[32] The constitutional amendment it took to do this happened without much kerfuffle.

Fui Asofou So'o points out that the Constitution that was agreed upon and enforced in 1962 gave Samoa 'a parliament [based] on the Westminster model, with separate arms for government' who make the laws, and for the judiciary who interpret or apply the laws.[33] Parliament, he says, 'comprises the head of state...and the legislative assembly'.[34] The office of the head of state is worth dwelling on as it underlines my point about the complex interplay of modern and traditional Samoan assumptions about leadership and the fluidity or blending that can sometimes happen between old and new ideas.

Today the head of state is elected by the legislative assembly. The term for a head of state is now 5 years. It is a position of leadership considered—even if only on ceremonial terms—to be the highest in the land. The parliamentary representatives who make up the legislative assembly and who elect the head of state are elected by two constituencies: 1. by their aiga (or extended families) in the first instance, in terms of their matai candidacy; and then 2. by their voting constituency under the universal suffrage regime, in terms of their political district candidacy. This means that aiga politics—the creation, consolidation and extension of old and new family loyalties—are still part and parcel of Samoan national politics and national leadership election processes.

To date the holders of the office of head of state have all been matai title-holders of considerable traditional rank and status within Samoa. To have been bestowed these titles meant that they had gained the support of their aiga (families), nuu

---

[29]See MNRE (2014, 2015).

[30]A significant proportion of that number rely for daily living and for funding family and community events or contributions (including monotaga) on remittances from family members living outside of Samoa (mainly in New Zealand, Australia and the United States of America). See Samoa Bureau of Statistics (2016) and Macpherson and Macpherson (2011).

[31]American Samoa is an unincorporated territory of the United States of America. It shares a strong and intimate cultural history with the independent state of Samoa.

[32]Angelo (2012), p. 145.

[33]So'o (2012), p. 44.

[34]Ibid.

(villages), and itumalo (districts), all of whom must agree to the bestowal. The first joint heads of state at independence were Tupua Tamasese Mea'ole and Malietoa Tanumafili II. Both were descendants and title holders of the highest ranked chiefly titles from the two leading aristocratic families of Samoa, the aiga Sa Tupua and the aiga Sa Malietoa. Their titles are tama-a-aiga titles. Eligibility to these tama-a-aiga titles and the leadership roles they entail was determined by Samoan custom, by the operation of both aganuu (customary principles common across Samoa) and agaifanua (customary principles particular to the place of their birth, especially to place/s from where their titles belonged).[35]

Both Tupua Tamasese Mea'ole and Malietoa Tanumafili II held their offices for life. Upon their deaths the current head of state, Tui Atua Tupua Tamasese Ta'isi Tupuola Tufuga Efi, who was elected by Parliament as per the Constitution, has now held the role for two terms and will be eligible for re-election for a further 5 years next year (2017).[36] He, like his predecessors holds a tama-a-aiga title, but as well, holds one of the four pāpā titles, that of the Tui Atua.[37] While according to the Constitution the position of head of state is not specifically the right of tama-a-aiga or pāpā title holders, in custom this is presumed. There is an ideological foundation to the tama-a-aiga and pāpā title systems that is hierarchical and culture-specific (but not gender-specific); a foundation built on philosophical ideas about divine designation (tofi), wisdom and balance (tofā and faautaga), and a sacred genealogical propinquity between all living things and God (implicit in ideas of tapu and va).[38] These ideas run against the grain of western human rights theories of leadership in parliamentary democracies where ideas of divine designation are considered repugnant and backward. But modern Samoan society and its modern state rely heavily on the divine ideologies of faakerisiano (Samoan Christianity) and faamatai (Samoan chieftainship) for definition and legitimacy. The question of whether the faamatai (to which the tama-a-aiga and pāpā title systems belong) and Parliament can co-exist is immaterial, because they are. The more pertinent question is when they clash—as seems the case with the current monotaga issue—what should we be asking first: which source of law ought to prevail? Or, where is the clash?

The statutory requirement that candidates for head of state do not need to be tama-a-aiga or pāpā titleholder but merely a member of the council of deputies seems to defy the tofi ideology inherent in the faamatai and faasamoa.[39] The only member of the current council of deputies that holds a tama-a-aiga title is Tuimalealiifano Vaaletoa Sualauvi II. Recently Samoa re-engaged in a public

---

[35] So'o (2008) and Tamasese (2016).

[36] The Constitution of the Independent State of Samoa 1960, Part III.

[37] The four pāpā titles are Tui Atua, Tui Aana, Gatoaitele and Vaetamasoalii. They are considered the four most paramount titles in Samoa. See So'o (2008) for a detailed explanation.

[38] See Suaalii-Sauni et al. (2009, 2014) for discussion on these concepts (tofi, tofā, faautaga, tapu and va).

[39] The Constitution of the Independent State of Samoa 1960, Part III, Art 25.

debate on whether the head of state and membership to the council of deputies ought to be reserved for tama-a-aiga or pāpā title holders only.[40] This was triggered by Prime Minister Tuilaepa's recent nomination of two non-tama-a-aiga/non-pāpā candidates (Le Mamea Ropati and Tuiloma Pule Lameko) for the council in February this year. Non tama-a-aiga or pāpā title holders have previously been members of the council of deputies, namely Vaai Kolone, Faumuina Anapapa and Matai'a Visesio, but none have yet progressed to the role of head of state. It will be interesting to see what will happen next year when the 5 year term of the current head of state ends and the election process begins again.

While, as Fui Asofou So'o and Jon Fraenkel pointed out in 2005, there is a sharpening of the divide between 'ceremonial customary authority and parliamentary leadership' in Samoa,[41] the matai and its faamatai continues to prevail nonetheless as a key organising social institution and system for Samoans, one that continues to have direct bearing on their everyday lives and their contemporary ideas about what are appropriate models of leadership for Samoa.

Samoa's current 50 seat legislative assembly was originally 45 seats before 1991. In 1991 a constitutional amendment allowed for the creation of two more seats increasing the number to 47, which was increased again in 1992 to 49. Of these 49, 45 were for the territorial constituencies and two for the individual voters (i.e. those who did not have rights to matai titles and customary lands and were registered on the individual voters roll). In 2013 the constitution was amended again to allow for women to have five or 10% of seats in Parliament. Because only four women were elected to Parliament in the 2016 elections the provisions of the 2013 constitutional amendment kicked in and so now Samoa's current 16th Parliament has 50 seats: five of them held by women, 45 by men; and four of these five women are members of the H.R.P.P party.[42] There have been 15 constitutional amendments since independence; 10 of which were passed from 1990 onwards.[43]

Except for the years between 1985 and 1987, the H.R.P.P party has enjoyed 30 years of more or less consecutive rule as Samoa's governing political party. They first came into power as Samoa's first political party in 1982 and have pretty much held the reins ever since. Their longevity is a first for any Pacific island state. After emphatically winning this year's election (they won 47 of the 50 seats), the H.R.P.P. seeks to extend their hold on Samoan independent nation state politics for another 5 years.

Samoan political commentator Fui Asofou So'o has argued that 'Samoan political parties have not been founded on any identifiable ideologies. Instead they have

---

[40]Tupufia (2016b) and Samoa Observer (2016). The first time this was debated was during the Constitutional debates. See Western Samoa Constitution Convention Debates, 1960.

[41]So'o and Fraenkel (2005), p. 356.

[42]Radio New Zealand (2016b).

[43]Mulitalo (2012b) and Angelo (2012). Both sources mention that between 1962 and 2012 there have been 14 amendments to the Constitution. The 15th amendment to the Constitution occurred in 2013. A copy of Mulitalo's unpublished work is on file with the author.

been a means of formally uniting members of Parliament in order to get the required number to form a government'.[44] So'o contends that 'there have been four main ways by which party unity' in Samoa has been achieved, 'with the H.R.P.P as the exemplar'.[45] He says: 'These are patron-client relationships, tightening of party rules, a conscious attempt to formulate and implement party policies, and the use of legislative mechanisms'.[46] 'Patron-client relationships', he says, is the mechanism most 'in tune with Samoa's family-oriented culture' and its 'cultural norms and practices' of 'reciprocity'.[47] Implicit in this 'family-oriented culture' and practices of reciprocity, or perhaps more precisely in aiga-oriented culture, are Samoan custom principles, those implicit in the ideologies of aiga, fa'asamoa, tu ma aga, aganuu and/or agaifanua.

Today the aiga, faamatai, faasamoa, aganuu and agaifanua concepts and institutions are assumed to inform and define Samoa's customary law system. However, few Samoans have a holistic and in-depth knowledge of that system. Most Samoan lawyers are trained in law schools outside of Samoa that privilege western frameworks of law and return to Samoa to practice in an environment that currently does the same. The internal logic of Samoan customary law principles and institutions have therefore remained underexplored, under-theorised and thus at risk of misunderstanding and misapplication. This kind of situation can open the door to wrongful judicial interpretations and unjust political manipulations of custom, of what it is and what it ought to be, and can therefore close the door to deeper investigations into how custom could and does benefit society. Unpacking the juristic elements of custom principles can help build the kind of juristic knowledge base needed for a Samoan indigenous jurisprudence. Let me illustrate this claim with reference to interpretations of the custom of monotaga as codified by the 2015 Electoral Amendment Bill No. 5, and interpreted by Samoan judges.

8.4   Monotaga and the 2015 Samoa Electoral Amendment Act No. 5

In 2015, the Samoan Parliament passed the Electoral Amendment Act No. 5 which defined monotaga in state law as:

> 'Monotaga' means the compulsory service, assistance or contribution (such as, contribution in form of cash, kind or goods) rendered for customary, traditional or religious activities, events, function or similar purposes pursuant to the customs of a particular village.

The terms 'service', 'village' and 'village service' are further defined as:

---

[44]So'o (2012), p. 71.
[45]Ibid.
[46]Ibid.
[47]Ibid.

'Service' means any service, assistance or contribution (such as, contribution in form of cash, kind or goods) rendered to a community for customary, traditional or religious activities, events, function or similar purposes.

'Village' is a village, from which a matai title was conferred, within a territorial constituency"; and

'Village service' is:

(a) monotaga rendered by a candidate in respect of one (1) or more of his or her matai titles within the territorial constituency in which the candidate intends to stand as a candidate; or

(b) service to a community by a candidate in an urban constituency.[48]

The rendering of this 'village service' or 'monotaga' must be 'for a period of at least three (3) years ending on the day in which the nomination paper is lodged with the Commissioner'.[49]

The Samoan version of the Act defines monotaga or village service in the following way. I offer an explanation of the Samoan terms I have emphasised in the Samoan text below to help underline my point about cultural nuance and the importance of having access to a deliberate conversation that offers rigorous analysis of the meanings behind any custom or indigenous concept and practice imported into state law. The Samoan definitions provided of monotaga and tautua (service) are direct translations of the English version of the Act and reads as follows:

'**monotaga**' o lona uiga o le **tautua e faamalosia**, fesoasoani po o saofaga (e pei o, saofaga i tinoitupe, ituaiga po o oloa) e tuuina atu mo galuega faaleaganuu, tu ma aga faaleatunuu po o le faale-lotu, mea e tutupu, faamoemoega faapitoa po o faamoemoega faapena, **e tusa ai ma aganuu a se nuu faapitoa**;

'**tautua**' o lona uiga so o se tautua, fesoasoani po o saofaga (e pei o, saofaga i tinoitupe, ituaiga po o oloa) ua tuuina atu i se nuu ma se afioaga i galuega faaleaganuu, tu ma aga faaleatunuu po o le faale-lotu, mea e tutupu, faamoemoega faapitoa po o faamoemoega faapena;

'**tautua i le nuu**' o lona uiga:

(a) **monotaga** tuuina atu e se sui faatu e tusa ai o sona igoa Matai e tasi (1) pe sili atu, i totonu o le itumalo faa-alaalafaga lea ua faamoemoe le sui faatu e tauva ai o se sui faatu; po o

(b) se **tautua** i se nuu ma se afioaga a se sui faatu i totonu o se itumalo o nuu tu taulaga....[50]

I draw your attention to the words 'tautua' and 'aganuu'. These two terms have a philosophical genealogy that can be traced to the 'ideologies of the previous order' (as per Sally Engle Merry). The term 'tautua', for example, is part of a well-known Samoan saying that is often quoted when talking about the requirements of a good Samoan leader. The saying goes: *o le ala i le pule le tautua* (literally: the path to

---

[48] See s 2. The Bill amends s. 3A of the principal Act. Available at http://www.palemene.ws/new/wp-content/uploads/Bills/2015/8/27%20August/Electoral-Amendment-Bill-No.-5-2015-Sam-Eng.pdf. Accessed 30 July 2016.

[49] Ibid, s 3A.

[50] Tulafono Tau Faaofi o Teuteuga o le Tulafono o Faiga Palota (Nu.5) 2015, 2, mo le fuaiupu 5 o le Tulafono autu.

leadership is through service).[51] But not just any type of service, it is about self-less service, the kind that privileges reciprocity and a selflessness that is assumed in the biblical text of Matthew 7:12, 'In everything, do to others what you would have them do to you' (NIV). It is the type of service assumed by the principle of alofa (which refers to ideas of agape: love and compassion). This understanding of tautua is one that cannot be made compulsory. To make tautua compulsory is to impose or demand something new of it. Such a change is repugnant not only to Samoan customary understandings of alofa, but also to contemporary Samoan Christian beliefs about the same. The question arises: what motivated the legislature to make such a change? Does the added requirement of mandatory-ness effectively create a new kind of monotaga, one not founded on customary law but on something else? If so, what is that something else?

When a matai title (and the pule or authority that goes with it) is bestowed on someone, it is because the family and village believes that he or she had given good enough tautua to warrant it. The tautua that is expected was never meant to be prescribed in fixed terms. It was always to be open to negotiation between those directly involved, even if in practice that may not be the case. This can be assumed by the privilege afforded the concept of alofa in the faasamoa. The common coupling of the principle of tautua and/or pule with alofa in Samoan custom underlines this point. Ideologically it is what offers balance in the search for wisdom (the tofā sa'ili).[52]

There are also different kinds of tautua. In Samoan custom three different kinds of tautua are recognised: tautua nofo tuāvae (service performed daily by someone that resides at the family household to the matai and the village), tautua 'ai taumalele (service given by somebody living outside the village), and osi āiga (service that is not regular but still considered important, such as presentations occasionally made to the family for family or aiga faalavelave).[53] Are these different types of tautua included in the meaning of tautua used to describe monotaga in the 2015 Electoral Amendment Act?

If we were explore tautua some more we would find that tautua was also service given by the untitled to the titled, not by the titled to those under his or her care. The service given by the titled to his or her family or village is usually described as tausiga.[54] The difference is subtle but important. The need to rename or distinguish the service or tautua of a chief, especially high ranking chiefs, links to a broader

---

[51] Suaalii-Sauni et al. (2009), Tamasese (2016), Suaalii-Sauni (2010) and Ngan-Woo (1985).

[52] Suaalii-Sauni et al. (2009).

[53] Suaalii-Sauni (2007), pp. 33–60.

[54] Ibid. Sister Vitolia Mo'a also explains that tausiga (or its root word 'tausi') is an ethical responsibility meaning 'to care for'. It is a sacred value that is all embracing and is implied in the role of a matai, in his/her responsibilities to his/her aiga (family), nuu (village), lotu (church) and atunuu (country). To this extent it is something inherent in the role and calling of a matai (chief). Tausi/tausiga, along with tautua, are part of a system of proportionality in the faasamoa that has as its primary desire or imperative to share in the carrying of burdens or duties (faamāmā avega). (Pers. comm.).

Samoan philosophy about the naming of things, from people and places to acts and values. The question here is where is the body of knowledge that can help us determine what this philosophy constitutes; its intellectual genealogy, hermeneutical foundations and juristic imperatives? To not know, or worse to refuse to want to know, is at best to plead ignorance, which is no defence in law (common law or custom law).

Samoan culture, the roles and responsibilities assumed within, is ordered and this is reflected in language. Just as the wisdom of the tulafale orator chief has a different name and purpose to that of the alii chief (i.e. the tofā mamao and the faautaga loloto respectively),[55] so too exists a difference in naming and defining untitled and titled service. Perhaps for some in law this is too obscure or subtle a point to be useful. I suggest not. Making sure lawmakers, especially legislative drafters, understand how best to capture nuance is critical when moving between languages, particularly when they carry inherently different worldviews. Indeed all involved in the lawmaking exercise, from the theoreticians to the practitioners, have a responsibility to ensure that they are well-versed in this nuancing exercise. If not, then whose responsibility is it?

The richness of meaning available on the concept tautua is also available for the concept pule, seen to be the likely outcome of good tautua. In Samoan customary law there are three different kinds of pule generally recognised across Samoa as part of Samoan aganuu. These are: pule faavae (constitutive authority), pule faasoa (distributive authority) and pule faamalumalu (protective authority).[56] There are also different kinds of matai, with different roles and responsibilities to their families and villages. There is an orator or tulafale matai and a high chief or alii matai. Some villages have combined these roles to form a hybrid matai, the tulafale-alii matai. The difference between matai is important because it determines the usual extent of their pule and other responsibilities, such as giving monotaga.

Everything in traditional Samoa was ordered or designated and this was based on theological presumptions about creation, relationships between living things, and with God. Such presumptions continue to govern everyday Samoan society today. God in the Samoan reference was not God distant creator. God was God progenitor, close ancestor.[57] This understanding of divinity nuances not only Samoan understandings of God and of the divine but also the faamatai, the faasamoa, aganuu and agaifanua. Can we assume this kind of rich, nuanced reading of the terms aganuu and tautua when interpreting their use in the 2015 Electoral Amendment Act?

Each family or village, according to their agaifanua or aganuu would establish a normative order for the rendering of tautua and for the tausiga they could expect. It

---

[55]Literally the words faautaga loloto or utaga loloto refer to the deep view or wisdom of an orator chief created through the type of burden or responsibility they carry. Loloto means deep; utaga refers to burden/responsibility. The terms tofā mamao means the wisdom of the long view or vision. Tofā refers to wisdom and mamao to long distance.

[56]Ibid. See also discussion on this by Vaai (1999).

[57]Suaalii-Sauni et al. (2009).

is within this context that the custom of monotaga sits. Like the subtle differentiation between tautua and tausiga it is of interest to note that no explicit legislative recognition was given to the concept of agaifanua in relation to monotaga in the statutory provision.[58] This raises the question, why the legislative preference for aganuu not agaifanua? Is agaifanua presumed like tausiga to be read within the terms aganuu and tautua respectively? What happens to the Samoan language, to Samoan customs and worldviews, if we did that?

The normative orders of aganuu and agaifanua are based on assessments of what is right and wrong that existed before the creation of Samoa as a nation state. Much of what existed before continues to inform understandings of aganuu today. Legislating for what is good tautua, the kind tantamount to giving support to a parliamentary candidate, finds the nation state directly imposing itself on what was once the prerogative of the village or aiga. It is here that arguments about the reality of legal pluralism in countries like Samoa are most pronounced. While the state, through the Village Fono Act 1990, recognises the custom law role of the village fono, it still assumes supremacy over the village fono and any custom laws it produces. This is evidenced by the statement in the Act that it is an Act 'to validate and empower the exercise of power and authority by Village Fono in accordance with the custom usage of their villages...'.[59] In other words, it is the state not custom or its non-state sources that validates and empowers the exercise of the power and authority of the Village Fono.

Monotaga as a cultural concept and practice has been around since at the least the 1800s, i.e. before European contact. Missionary linguist Reverend George Pratt recorded it as referring to 'the portion of food prepared by an individual to take with all the rest of the village to a party of visitors, o lana monotaga'.[60] It is translated by Fui Asofou So'o to mean in the contemporary context: 'traditional contributions to the village or participation in family social obligations'.[61] In both instances monotaga is about giving a type of service or contribution, one considered traditional in origins, and relating to the gifting of something to family and village as a show of solidarity and hospitality. Monotaga in this sense is part of a collective normative culture of responsibility, used to assist the village (nuu) and family (aiga) institutions to function well and peacefully. On this front monotaga may be better considered an ethical and cultural responsibility than a legal one.

All of this rich nuanced cultural context comes alive when close attention is paid to the Samoan not the English explanations of monotaga. When the English is emphasised the difference between a tulafale and an alii matai, between the various kinds of tautua and pule, between tautua and tausiga, etc., are easily lost or sidelined. When comparing the two language versions tell-tale signs that the Samoan statute is a translation of the English, rather than the other way around,

---

[58]Electoral Amendment Act (No. 5) 2015.

[59]Village Fono Act 1990. The Village Fono Amendment Bill 2015 does not remove this wording.

[60]Pratt (1893), p. 224.

[61]So'o (2007), p. 261.

emerge. The most telling sign is the translation of the idiomatic English concept of 'in kind', suggested in the phrase 'contribution in form of cash, *kind* or goods'.[62] When one reads the Samoan translation: '...saofaga i tinoitupe, *ituaiga*, po o oloa',[63] one finds a literal translation of 'kind' as 'type' rather than 'kind' as 'in kind'. The difference is subtle but like tautua/tausiga is important. In the English version of the statute the meaning 'in kind' can be assumed by the words around it. This is not possible in Samoan and so without the English the Samoan use of the term 'ituaiga' becomes ambiguous. I turn now to two Supreme Court cases for further insight into how we might make sense of the custom concept monotaga, or at least of how the judiciary are making sense of it.

8.5 Malielegaoi v Tuitui, and Samu v Adams

The Samoa Courts have been reluctant to depart from the text of the law and from previous rulings in cases that have tested the constitutionality of the 1963 Samoa Electoral Act and its various amendments. This was the situation for the *Le Tagaloa Pita* case which was an appeal against the Electoral Amendment Act 1990, as well as for *Mulitalo v Attorney-General* case, two of the more well-known and applied cases in this area.[64] It was in the 1995 *Le Tagaloa Pita* decision that the Court of Appeal declared that where there are differences between the English and Samoan versions of a law, including the supreme law, the English version is to prevail.[65] When the argument was raised about an incorrect language translation, the Court in that case resorted to the wisdom of the doctrine of precedent and argued: 'why, if the translation was incorrect, it had not been raised previously'? Such reasoning creates barriers to correcting translations that are clearly wrong or no longer applicable and likely to cause grave injustice.

Leading up to the 2016 elections the Prime Minister took a rival to Court claiming that he had not satisfied the monotaga requirements of the Electoral Act.[66] The respondent Tuitui opposed the Prime Minister's motion on the grounds that: 1. the Electoral Act 1963 was unconstitutional (i.e. in violation of Article 15, which provides for freedom from discriminatory legislation). He argued in particular that the 'different residential requirements for candidates for a territorial constituency being 3 years and for a candidate for an urban constituency being 6 months is discriminatory'. And, 2. that he had 'satisfied the village service

---

[62]Emphasis added.

[63]As above.

[64]See reference to these cases in Angelo (2012).

[65]Angelo (2012), p. 153.

[66]*Malielegaoi v Tuitui* [2016] WSSC 5. The case was presided by Chief Justice Sapolu and Justice Vaai.

requirements by contributing to the affairs and fundraising activities of his village such as raffles for building the village church and other financial activities'.[67]

The Court (made up of two Samoan judges) reasoned that: on the evidence the respondent had not fulfilled the 3 year village service requirement because, among other things: 'The electrical services which he carried out to the homes of some of the people of his village were paid for and cannot be described as monotaga. Likewise, the electrical wiring services that the respondent carried out to the Saleapaga primary school building in 2011 for which he sent a bill of costs to the village to pay cannot be described as monotaga'.[68] Therefore, the Court was satisfied on the evidence that the respondent had 'not carried out any monotaga [as per the Act] to his village of Saleapaga pursuant to the customs of the village, at least for the last 3 years prior to lodging his nomination paper'.[69]

After considering the evidence, case law and the constitutional provisions, the Court, applying *Mulitalo v Attorney General,* declared in favour of the applicant, in this case the Prime Minister. Tuitui was unable to continue his candidacy and the Prime Minster was elected unopposed. From this case we start to get a picture of what the Courts consider part of the statutorily-defined monotaga or village service. In this case the idea of tautua as selfless service as understood in custom is upheld as recipients should not have to pay for tautua, although this was not part of the Court's reasoning.

In the second case of *Samu v Adams,*[70] also heard by the Supreme Court and by Samoan judges, the Court provides further discussion on what monotaga might mean in terms of the statute. The motion before the Court focused on the claim that the respondent had not 'satisfied the village service [or monotaga] requirement' of the Act. That is, the applicant (Samu) argued that the respondent (Adams) attended the Lalomanu village fono for the first time on the 5th of April 2014 and so his candidacy for Parliament was therefore invalid as per the amendment.

The respondent counter-argued stating that while 'it was his first attendance of a village fono...he was bestowed the matai title in 1984 and since then, through the head matai (sa'o) of his family, Letalu Lokeni, [he had] rendered services to his village'.[71] He admitted though that, among other things, 'when the sa'o [head matai] of his family died in 2005, [his] services to his family, village and church then lapsed. [Moreover, he] was also absent from Samoa in 2006 and 2007 for further studies in New Zealand' during which time he did not continue his contributions.[72]

---

[67]Ibid, para 10.

[68]Ibid, para 16.

[69]Ibid.

[70][2016] WSSC 10. This case was presided over by Chief Justice Sapolu and Justice Vaai.

[71]Ibid, para 11.

[72]Ibid, para 12.

Like the *Tuilaepa v Tuitui* case, the respondent (Adams) had offered professional services to his village and had counted this as part of his monotaga or village service under the Act. The village service in question occurred when the tsunami hit Samoa in 2009 and continued thereafter. However, Dr. Adams declared that even though his patients had to pay only five tala to the Lalomanu hospital, they did not have to pay for his service fee. In his affidavit he stated:

> In all the circumstances it is difficult to put a monetary sum on my total tautua, the quality and the quantity, suffice to say that I have provided medical consultation services for free to just about every resident of Lalomanu village, from the elderly to a new born baby during this period, October 2009 to May 2015 [at least 6 years].[73]

The applicant counter-argued stating that the free medical services that Dr. Adams referred to were in fact not free as the respondent was paid 150 tala an hour by the National Health Service to run his weekly clinics in Lalomanu during this time. Interestingly and somewhat confusingly the Court found that on their reading of the evidence it is satisfied that the respondent had rendered service to the village for the requisite 3 year period. It stated:

> The court accepts that in April 2014 the respondent commenced to attend the village fono monthly meetings and personally rendered his monotaga to the village rather than through the sa'o of his family. Which means that when his weekly clinics ceased in May 2015 the respondent was still rendering service through his monotaga till the day of nomination. *It follows that the 3 year period prior to nomination is not an issue.*[74]

Leaving aside whether there was a typographical error in the above paragraph for the moment (i.e. that the last sentence should have read 'is an issue' rather than 'is not an issue'), or that the 3 year period refers to something other than the compulsory period for rendering monotaga, what the Court then, somewhat clumsily, focuses on is whether the service or monotaga rendered was monotaga as prescribed under the Act. The judgement centred on the Court's interpretation of the medical services provided by the respondent from October 2009 to 2015. Did it fit within section 5 (3A) of the Act or not? The Court reasoned that it did not. The Court stated:

> The court agrees with counsel for the applicant that the medical service provided by the respondent cannot be termed monotaga as it was not a compulsory service, assistance or contribution, rendered for the customary, traditional or religious activities pursuant to the customs and traditions of Lalomanu. The weekly medical clinics were arranged by the National Health Service of Samoa who paid the respondent $150 per hour to manage the clinic.[75]

The Court went on to say:

> It was not until April 2014 that he [the respondent] recommenced rendering monotaga by personally contributing, and rendering service to the customary and traditional activities of

---

[73][2016] WSSC 10. This case was presided over by Chief Justice Sapolu and Justice Vaai, para 13.
[74]Ibid, para 19 (emphasis added).
[75]Ibid, para 20.

the village. His compulsory service to the village through his monotaga therefore did not commence until April 2014, less than 3 years prior to close of nomination".[76]

Therefore, the Court concluded, 'He has obviously not satisfied the village service requirement.'[77]

The reasoning employed here is contradictory. The Court seemed to favour an interpretation of monotaga as constituting, 1. compulsory service to the village; and 2. given for customary, traditional or religious activities pursuant to the customs and traditions of Lalomanu. Meeting these two criteria depended ultimately on whether Dr. Adams' weekly clinics can be counted as monotaga or not. The Court decided it could not because Dr. Adams was being paid by the NHS for this service. Unfortunately, there were no submissions recorded relating to what constituted Lalomanu village customary and traditional activities or expected monotaga. Interestingly the Court underlined the qualifying words 'compulsory' in its recitation of section 5 (3A) of the Act in paragraph 20 of its judgement. Although the Court did not elaborate beyond underlining it, it is, for reasons outlined earlier, an element worth interrogating for what that says about the Court's position on monotaga as compulsory.

What these cases illustrate is the tension between the flexibility of village custom to define and adapt itself to its own context irrespective of what other villages are doing (which is the basis of agaifanua), and the prescriptive and standardising nature of statutory law on the one hand, and aganuu—where it prescribes a norm for all—on the other. At the time of writing Le Tagaloa Pita, a candidate from the village of Sili, was disqualified for not giving monotaga to his village pursuant to the definition of monotaga under the Electoral Act. It is well known in Samoa that according to the customs of Sili an alii chief of significant status does not give monotaga. They give in other ways but these are not described as monotaga. Le Tagaloa is quoted in the Samoan Observer as saying: 'The decision is discriminating and against the customs and traditions of my village and others that might be the same. . . .the law is basically forcing our village to go against its customs'.[78] It will be interesting to note how the Court will rule on this one if Le Tagaloa is able and decides to appeal the decision.

The judiciary play a critical role in determining the true place of custom in law. But so too do legislative drafters, legal practitioners and socio-legal theorists. It is crucial that at this stage in Samoa's legal development that all these parties come together to build a system capable of honouring the long unfulfilled promise of its constitutional drafters. This brief but close analysis of monotaga highlights just how complex custom in law is, but with a bit of dedicated time and professional and political will some real inroads can be made to ensure that the best of it remains alive and well. Dr. Adams was understandably frustrated by the Court's decision in his case and when he was asked by a reporter about how he felt about it, he called

---

[76][2016] WSSC 10. This case was presided over by Chief Justice Sapolu and Justice Vaai, para 23.
[77]Ibid, para 24.
[78]Tupufia (2016c).

for a review of the Act.[79] A review will not make much difference, however, not without the resources to develop a jurisprudence for Samoa that judges will take seriously; a jurisprudence that speaks intelligently to the intellectual and philosophical premises of custom law and its legitimacy as a key source of law for Samoa.

8.6   Concluding Reflections

In this chapter I conducted a close examination of the concept and custom of monotaga in order to offer insight into the rich texture of legal pluralism in Samoa. The chapter offers some appreciation of the difficulties implicit in defining laws when the meanings associated with them derive from multiple sources. I argued that custom is an intricate part of Samoa's contemporary socio-political fabric; it is dynamic and malleable. Assessing custom's relevance requires having a body of scholarship available to inform a debate on law and its ideological values, core institutions, practices and principles, such as monotaga, aganuu, agaifanua, faamatai, tautua, alofa and pule.

While there is a rich body of scholarship within legal pluralism, legal anthropology, sociology and Pacific studies that has begun the work of unpacking custom for its juristic qualities, there is still some way to go. Huffer and So'o in their analysis of Samoan political thought acknowledge that custom continues to play a central role in Samoan life, and conclude that 'greater critical awareness and public debate of and engagement with the norms and practices of both systems' is required.[80] They re-emphasised the call by Huffer and Qalo (2004) to Pacific academics to establish a body of scholarship that can offer 'a genuine and far-reaching contextualization...of indigenous cultural values in contemporary settings'.[81] This chapter seeks to respond to that call.

In the field of Pacific Studies, Epeli Hau'ofa's (1994) essay 'Our Sea of Islands' is considered seminal work.[82] His essay was borne out of a concern for the profound limitations that taking a 'size' approach can have on a person's spirit. Seeing first-hand its negative effects on his many students from small developing Pacific island nations, Hau'ofa advocated for a reimagining of Oceanian indigenous institutions and values and a re-articulating of their relevance to contemporary society. Like the dream of a South Pacific jurisprudence,[83] the dream implicit in Hau'ofa's reimagining requires, for any level of achievement, solid intellectual *and* political commitment.

---

[79] Ale (2016).

[80] Huffer and So'o (2005), p. 327.

[81] Huffer and Qalo (2004), p. 108.

[82] Hau'ofa (1994).

[83] Corrin (2016b). Published in Professor Corrin's abstract in the conference programme.

Pacific island national elections offer a useful case study to analyse some of the complex, conflicting and intersecting boundaries between law and politics. Given the very complex and highly politically charged environment in which the Electoral Act operates, judges in small island one-party states like Samoa have the unenviable responsibility of upholding the rule of law and keeping the legislature honest, whether they like it or not. This is a lot of responsibility to put on the shoulders of just one group in society and is itself not without power issues.

For legal scholars committed to the positive development of the Pacific region, delving into custom law archives and making it intelligible to the legal and cultural theorist and practitioner alike is, I contend, a must. We do it not to essentialise or romanticise custom but to find points for comparison, to show contemporary relevance, and to highlight historical continuities and discontinuities so that law can indeed be for all. Without the development of this kind of juristic work, we will not be able to even begin to address the promise of what former Fiji Family Law Court Judge, Madam Mere Pulea, and Professor Jennifer Corrin allude to in their work on justice and jurisprudence in the Pacific. Using some of their words I end this chapter by musing over and imagining a conversation between them and myself. The conversation goes:

> Professor Corrin: "[You know], although much lip service has been paid to local culture and traditional values, why has the dream of a South Pacific jurisprudence, which was expressed so eloquently in the preambles to many new constitutions, not translated into reality?"[84]
>
> Judge Mere, responds: "But what kind of justice are we searching for? One that is of 'high quality' in the sense of rigid conformity to English legal practices and values? Or do we seek the kind of judgements that are firmly rooted in the Pacific context where judges are attuned to the customs, conditions and way of life of the people they are judging?"[85]
>
> But I, drawing on Natalie Baird,[86] think to myself: "And who will judge our judges?!"

References

Aiono-Le Tagaloa F (2009) Sua le lea, toto le ata: the land and titles court of Samoa, 1903–2008: amid continuity and change. PhD Thesis, Faculty of Law at the University of Otago

Ale PM (2016) Disqualified candidate wants Act reviewed. In: Samoa Observer. Available at http://www.samoaobserver.ws/en/27_02_2016/local/2998/Disqualified-candidate-wants-Act-reviewed.htm. Accessed 30 July 2016

Angelo AH (2012) Steady as she goes: the constitution and the court of appeal of Samoa. NZACL Yearb 18:145–165

Baird N (2014) Judges as cultural outsiders: exploring the expatriate model of judging in the Pacific. Canterbury Law Rev 19:80–96

Brown K (1999) Customary law in the Pacific: an endangered species? J S Pac Law 3. Available at https://www.usp.ac.fj/index.php?id=13145. Accessed 30 July 2016

---

[84]Ibid.

[85]Baird (2014), p. 80.

[86]Ibid.

Corrin J (2009) Moving beyond the hierarchical approach to legal pluralism in the South Pacific. J Leg Pluralism 59:29–48

Corrin J (2016a) Dispelling the myths of legal pluralism. Oral Presentation given at the Pasifika Law and Culture Conference, Victoria University of Wellington

Corrin J (2016b) Pacific legal systems: past, present and future. Panel presentation as part of the 6th Biennial Conference of the Australian Association of Pacific Studies, Pacific Pasts - Pacific Futures, James Cook University

Foucault M (1979) Discipline and punish: the birth of the prison. Vintage, New York

Foucault M (1994) Governmentality. In: Rabinow P, Rose N (eds) The essential Foucault: selections from essential works of Foucault. The New Press, New York, pp 1954–1984

Harris P (2002) An introduction to law, 6th edn. Butterworths Tolley

Hau'ofa E (1994) Our sea of islands. Contemp Pac 6(1):147–161

Huffer E, Qalo R (2004) Have we been thinking upside-down? The contemporary emergence of Pacific theoretical thought. Contemp Pac 16(1):87–116

Huffer E, So'o A (2005) Beyond governance in Sāmoa: understanding Samoan political thought. Contemp Pac 17(2):311–333

Keresoma L (2013) Prime minister wanted Lafaitele expelled from party. In: Talamua Online News. Available at http://www.talamua.com/prime-minister-wanted-lafaitele-expelled-party. Accessed 30 July 2016

Macpherson C, Macpherson L (2011) Churches and the economy of Samoa. Contemp Pac 23 (2):304–337

Marck J (1996) The first-order anthropomorphic gods of Polynesia. J Polyn Soc 105(2):217–258

McLeod I (2007) Legal theory, 4th edn. Palgrave MacMillan, New York

Meleisea M et al (1987) Lagaga: a short history of western Samoa. University of the South Pacific, Fiji

Merry SE (1988) Legal pluralism. Law Soc 22(5):869–896

MNRE (2014) Island biodiversity national theme: embracing the importance of Samoa's biodiversity. Available at http://www.mnre.gov.ws/index.php/biodiversity/83-environment-a-con servation. Accessed 30 July 2016

MNRE (2015) Aleipata communities support management of invasive species at Aleipata islands. Available at http://www.mnre.gov.ws/images/stories/MNRE/DEC/AleipataIslandsConsulation. pdf. Accessed 30 July 2016

Mulitalo TL (2012a) The practice of legislative drafting in Samoa, a plural society of the South Pacific. Loophole Commonw Assoc Legislat Couns 3:28–44

Mulitalo TL (2012b) The constitution of the Independent State of Samoa: constitutional amendments 50 years on, 1962–2012. Unpublished work. Copy available with the author

Mulitalo TL (2013) The consequences of legal pluralism for law reform in the South Pacific. PhD Thesis, TC Bierne School of Law at the University of Queensland. Available at http://espace. library.uq.edu.au/view/UQ:314962. Accessed 30 July 2016

New Zealand Law Commission (2006) Converging currents: custom and human rights in the Pacific. In: Law Commission study paper 17. Available at http://www.lawcom.govt.nz/sites/ default/files/projectAvailableFormats/NZLC%20SP17.pdf. Accessed 30 July 2016

Ngan-Woo F (1985) Faasamoa: the world of the Samoans. Office of the Race Relations Conciliator, Auckland

Pratt G (1893) Pratt's grammar dictionary Samoan-English English-Samoan, 3rd edn. London Missionary Society, London

Radio New Zealand (2016a) HRPP 47 seats, Tautua three - official result from Samoa election. Available at http://www.radionz.co.nz/international/pacific-news/298738/hrpp-47-seats,- tautua-three-official-result-from-samoa-election. Accessed 30 July 2016

Radio New Zealand (2016b) Samoa's ruling party given resounding mandate in elections. Available at http://www.radionz.co.nz/international/pacific-news/298183/samoa's-ruling-party- given-resounding-mandate-in-elections. Accessed 30 July 2016

Samoa Bureau of Statistics (2016) Population & demography indicator summary. Available at http://www.sbs.gov.ws/index.php/population-demography-and-vital-statistics.   Accessed 10 April 2016

Samoa Observer (2016) Party politics claim denied. Available at http://www.samoaobserver.ws/en/25_02_2016/local/2906/Party-politics-claim-denied.htm. Accessed 30 July 2016

Simmonds NE (2008) Central issues in jurisprudence, 3rd edn. Sweet & Maxwell, London

Suaalii-Sauni T (2007) E faigatā le alofa: the Samoan fa'amatai – reflections from afar. In: So'o A (ed) Changes in the fa'amatai system: o suiga i le fa'amatai. National University of Samoa, Apia

Suaalii-Sauni T et al (eds) (2009) Su'esu'e Manogi: In search of fragrance: Tui Atua Tupua Tamasese Ta'isi and the Samoan indigenous reference. Centre for Samoan Studies at the National University of Samoa, Samoa

Suaalii-Sauni T (2010) Samoan custom and discourses of certainty. In: Benton RA (ed) Yearbook of New Zealand Jurisprudence, vols 13–14. University of Waikato, Hamilton

Suaalii-Sauni T et al (eds) (2014) Whispers and vanities: Samoan indigenous knowledge and religion. Huia Publishers, Wellington

So'o A, Fraenkel J (2005) The role of ballot chiefs (matai palota) and political parties in Samoa's shift to universal suffrage. Commonw Comp Polit 43(3):333–361

So'o A (ed) (2007) Changes in the fa'amatai system: o suiga i le fa'amatai. National University of Samoa, Apia

So'o A (2008) Democracy & custom in Sāmoa: an uneasy alliance. IPS Publications, Fiji

So'o A (2012) Political development: Samoa's parliamentary journey from 1962 to 2012. In: Meleisea M et al (eds) Samoa's journey 1962–2012: aspects of history. Victoria University Press, Wellington, pp 44–76

Tamasese TAT (2016) Keynote address: where is our island? Navigating language, vision and divine designation in Samoan law and jurisprudence. Keynote Address, Samoa Law Society & Te Hunga Rōia Māori o Aotearoa Joint Conference, 7 July 2016, Apia, Samoa. Available at http://www.headofstate.ws/where-is-our-island-navigating-language-vision-and-divine-desig nation-in-samoan-law-and-jurisprudence. Accessed 30 July 2016

Tupufia LT (2016a) Associate minister wanted. In: Samoan Observer. Available at http://www.samoaobserver.ws/en/03_04_2016/local/4494/Associate-Minister-wanted.htm.   Accessed 30 July 2016

Tupufia LT (2016b) Savai'i ignored, Tautua claims. In: Samoa Observer. Available at http://www.samoaobserver.ws/en/04_02_2016/local/2004/Savai'i-ignored-Tautua-claims.htm.   Accessed 30 July 2016

Tupufia LT (2016c) Faumuina successful, Le Tagaloa disqualified. In: Samoan Observer. Available at http://www.samoaobserver.ws/en/26_02_2016/local/2948/Faumuina-successful-Le-Tagaloa-disqualified.htm. Accessed 03 Sept 2016

Vaai S (1999) Samoa faamatai and the rule of law. National University of Samoa, Samoa

Part III
The Legal Profession in Small States: Education, Practice and Regulation

Chapter 9
Legal Education and the Profession in Three Mixed/Micro Jurisdictions: Malta, Jersey, and Seychelles

Seán Patrick Donlan, David Marrani, Mathilda Twomey, and David Edward Zammit

9.1 Introduction

Among the many interesting research problems of a general nature found within the area of comparative law, two complex issues are particularly fascinating. . .: the problem of mixed (hybrid) legal systems and the problem for the so-called "small jurisdictions".—Bogdan (1989)

Even when geographically isolated, the small populations—both general and legal—of micro-jurisdictions and small states typically require them to reach beyond, often far beyond, their shores. This doesn't mean that they are isolated intellectually. Instead, their necessary engagement with outside traditions and influences makes them crossroads of interaction between peoples, jurisdictions, and powers. This is especially true of so-called mixed jurisdictions where the legal tradition is an explicit, obvious hybrid. With these encounters in mind, this chapter explores legal education and training and the legal profession in three mixed/micro-jurisdictions: Malta, Jersey, and Seychelles. The three provide a useful focus for comparative analysis. Each jurisdiction is an island or archipelago. Both Malta and

S.P. Donlan (✉)
School of Law, University of the South Pacific, Port Vila, Vanuatu
e-mail: sean.donlan@vanuatu.usp.ac.fj

D. Marrani
Institute of Law Jersey, St Helier, Jersey
e-mail: david.marrani@lawinstitute.ac.je

M. Twomey
Supreme Court of Seychelles, Mahe, Seychelles
e-mail: m.twomey@judiciary.gov.sc

D.E. Zammit
Faculty of Laws, University of Malta, Msida, Malta
e-mail: david.zammit@um.edu.mt

© Springer International Publishing AG 2017
P. Butler, C. Morris (eds.), *Small States in a Legal World*,
The World of Small States 1, DOI 10.1007/978-3-319-39366-7_9

Jersey are within Europe, while Seychelles is a considerable distance away geographically, and arguably culturally as well. Jersey and Seychelles are particularly
small jurisdictions with populations of less than 105,000 each. By comparison,
Malta is significantly larger, with more than four times that number of people,
though it is still well within common definitions of micro-jurisdictions. It is not
unusual, for example, to limit that category to systems with fewer than 1,500,000
people. The legal traditions of both Jersey and Seychelles have important French
influences. Malta has long had much in common with its Italian neighbour, both
before and after the Risorgimento of the nineteenth century. As with much of
Europe, it also drew inspiration from French developments in the same period.
And Malta, Jersey, and Seychelles each have significant links, in both the past and
the present, with Britain. The result is that comparison of these traditions provides
us with the opportunity of observing their different responses to legal education and
to the legal profession.

Our intention is to (1) broadly lay out the patterns of legal education and training
and the legal profession in these jurisdictions and (2) briefly examine the degree to
which legal education and training and the legal profession in them is dependent on
external influences. For the former, the three systems have adopted related, but
distinct approaches to education and training that are usefully compared. We also
consider how insiders in these jurisdictions look abroad to orient their studies and
practice and investigate the internal use of legal actors and sources from outside:
jurists and doctrine, judges and jurisprudence, legislators and legislation, as well as
foreign-trained practitioners. The effect of such external influences in small jurisdictions is profound, perhaps especially in explicitly mixed traditions. Indeed, in the
words of Sue Farran, such influences—she was writing specifically about the provision of legal education by outside institutions in foreign and hegemonic and
neo-colonial traditions—have the potential to act as 'trojan horses' for a sort of
'legal imperialism'.[1] Inevitably, our discussion will touch, if only lightly, on questions not only of self-government—indeed of legal purity, pragmatism, and
pollutionism—but even of self-identity.[2]

9.2 Malta

The Republic of Malta (Malta) consists of three islands: Malta, Gozo, and Comino.
They are administered collectively, have a combined population of 446,500, and
cover only some 300 km^2. Sicily sits less than 100 km to the north and has long

---

[1]Farran (2013a), p. 350. See also the related plenary address from Farran (2013b).

[2]In discussing specific hybrids of European laws, Palmer (2012) refers to purists who want to
protect the original layer of continental law, pollutionists who want to add more Anglo-American
law (to the existing layer of such laws), and pragmatists who aren't committed either way, but who
require reasonably clear answers to legal issues that arise in the jurisdiction.

played an important part in Maltese history. But Tunisia to the west and Libya to the south are not much farther away. Malta has a colourful past. Linked to Sicily until 1530, when it was gifted to the Knights of Malta by Emperor Charles V, the Knights were in turn expelled by Napoleon in 1798. The British arrived 2 years later, having helped the Maltese rebels evict the French. By 1815, the archipelago was effectively incorporated into the British Empire. The result was an essentially colonial relationship, where the island of Malta was utilised as a fortress and port for the British Empire from 1800–1979.[3]

Maltese colonialism was different, of course, from forms taken beyond Europe. Since Malta was already a densely populated island with a European culture, there was no sustained attempt to settle a large British population. And after initial confrontations with the Maltese legal elite about the language of the laws, the British largely accepted the continued operation of the Courts, staffed overwhelmingly by Maltese lawyers and judges, speaking Italian and administering a law rooted in the ius commune. Some of the Maltese leadership actively solicited British rule, expecting them to act as "il piu" paterno dei governi'.[4] Up to the 1940s, nationalists looked towards Italian language and culture as a source of identity and inspiration. Indeed, the so-called 'Language Question' dominated politics between the 1870s–1930s, revolving around whether Italian or English was to be the language of education, law, and administration. Maltese politics developed within this context, as did the eventual recognition of Maltese as an official language (rather than just an oral dialect). When, in the 1930s the British government changed the language of the laws and legal education from Italian to English, an Italianised Maltese remained the language of the Courts, the laws, and state administration. English remains widespread. Legislation is published in both Maltese and English; but where there is conflict between texts, the Maltese usually prevails. Italian is still widely known, but has no official status as such.

An inclination towards one or another external cultural and legal influences—Britain or Italian—has a long genealogy in the context of Maltese state building. In the 1950s, Malta's Labour Party argued for full political integration into the United Kingdom (UK). They also made secret plans to integrate Malta into Italy if this failed. Eventually, after integration with the UK was rejected in a referendum, the party advocated complete independence for Malta, coupled with strict neutrality in foreign affairs and closure of the British military base in Malta. The Nationalist Party advocated political independence, which was obtained in 1964; Malta remains a member of the Commonwealth of Nations. The Party subsequently advocated European Union (EU) membership. This was achieved in 2004.

---

[3]See Donlan et al. (2012) for an extended discussion of Maltese legal mixity. Much of the text on Malta draws from this article, though it is updated to reflect changes since 2012. See also Andò et al. (2012).

[4]'They (the Maltese leaders) expected Britain to safeguard their interests, to revive the economy and to grant some form of representative rights. But essentially they expected Britain to act as "'il piu' paterno dei governi" ["The most paternal of governments."] as they recognised the British king as their new sovereign'. Translation from Zammit (1984), p. 10.

Membership is seen, consistent with this history, as compatible with the exercise of Maltese sovereignty. Moreover, EU membership reconciled the various loyalties—British, Italian, and Maltese—adhered to in Malta. As a result, it was possible to be all three at once.

Malta is thus, with the Bailiwicks of Guernsey and Jersey, Cyprus, and Scotland, one of a handful of explicitly mixed legal systems within Europe. The jurisdiction combines a fundamentally continental civil law showing pervasive French and Italian influence with a modern commercial law which is predominantly borrowed from British law.[5] Delicts or torts has been significantly influenced by Anglo-British models. The trust has also been introduced, though relatively recently. Succession law has largely survived. Over time, especially since the mid-twentieth century, first legislation, then judicial development, brought about a major shift towards British models. Similarly, Maltese substantive criminal law blends continental traditions, especially the Italian, with British influence. Public law is more clearly British in both form and substance. Maltese procedures are mixed, too. Civil procedures are significantly judge-centred or 'investigative' in the continental manner, while criminal procedures owe more to British law. The entire mix also owes much to European law (of both the EU and the European Court of Human Rights (ECtHR)) and international law.[6] Much of its legislation, both substantive and procedural, is hybrid in character and, as with other explicitly mixed systems, Malta suggests the significant limitations of traditional comparative taxonomies.

These complexities extend beyond generalities to Maltese legal sources and legal practice. In addition to European law, its enacted law consists of several hundred 'chapters' arranged chronologically. The Constitution itself is quite long and arguably reflects the precision and detail of English statutory drafting. As in continental jurisdictions, however, Malta's written law includes continental-style codes (in contrast to mere restatements of law) as well as narrower, special legislation. Custom remains a binding source of law, but its role is fairly limited in practice. Even Roman law may occasionally be relevant as a direct source of Maltese law in its own right. While not an official source of law, Maltese jurisprudence carries greater weight than in continental jurisdictions. That is, its case-law is formally persuasive and non-binding rather than binding in the manner of Anglo-American stare decisis. But Maltese jurisprudence reflects other elements associated with Anglo-American judicial methods, e.g. the ratio decidendi/obiter dictum distinction may be invoked. Indeed, as in Anglo-American jurisdictions, the judiciary play a greater role in developing its law than do legal scholars. Especially since the end of the Second World War, the opinions of Maltese judges have increasingly resembled more the discursive pattern of the Anglo-American law than the more abbreviated form associated with the Franco-Romano (rather than Germanic) continental tradition. This reflects a wider Anglophone influence, but it is not total. As with continental practice, the judiciary issues collegial opinions,

---

[5]Castellucci (2014).

[6]See generally Donlan et al. (2012); Aquilina (2011), p. 261.

individual judges are not identified, and dissents are not issued. Codal and statutory interpretation can follow either continental and Anglo-American approaches, whichever is appropriate to the texts involved. And, unlike British and Irish practice, 'parliamentary history' is invoked in Maltese Courts.

Unlike Jersey and Seychelles, Maltese legal education and training, both academic and professional, is entirely domestic. Its only law faculty is based in the University of Malta. The faculty consists largely of part-time lecturers simultaneously engaged in legal practice, but full-time lecturers are increasingly common. The language of instruction is mainly English, not least to accommodate a significant number of foreign students in its law programme. At the same time, Maltese words may sometimes be used in the teaching of procedural law, where an English equivalent may not always exist. This is important as Court practice and jurisprudence are conducted almost wholly in Maltese.[7] Instruction occurs largely by lectures and tutorials, although courses based on experiential learning are now being developed in the context of a new Masters level course in Advocacy.

Legal education combines the reading of both foreign and domestic texts. Students are expected to be familiar with Court judgements delivered in Maltese, as well as with law books written in English and Italian, depending on the topic. Domestic texts include the lecture notes of leading (past) jurists and a developing body of locally-produced scholarship. Like legal education across the continent, law students complete an undergraduate degree in law followed by a graduate degree required for legal practice. Until recently, these degrees were the LLB and the LLD respectively. While these degrees developed from the Anglophone or British legal/academic tradition, Maltese study differed from (i) the British and Irish combination of university followed by professional study outside of the university and (ii) the three-year graduate study of law in North American law schools. During their LLD study, students would be apprenticed with practitioners.

In the last few years, a number of changes have occurred to bring Malta in line with the Bologna Process of the EU. That process was meant to harmonise, to a degree, legal education across Europe to facilitate movement across its boundaries and legal systems. The resulting reforms in Malta mean that students will study for less time than in the past. From 2017, the overall duration of their law studies has been reduced to 5 years, consisting of a four-year long LLB followed by a 1 year Masters course in Advocacy or Notarial Studies. And whereas past students produced lengthy academic dissertations of 35,000 words for the LLD, students now complete a much shorter, 10,000 word research paper. In addition, while Maltese advocates traditionally used 'Doctor' as their academic and professional title, that practice will, in future, be purely conventional and informal.

The required first year classes at the Faculty of Laws include:

1. Basic Notions of Commercial Law
2. Comparative Legal Systems & Legal Pluralism

---

[7]Admission to legal study requires high levels of facility in both Maltese and English.

3. Constitutional Law
4. Introduction to European Union Law
5. Introduction to Law
6. Introduction to Legal History
7. Philosophy of Law
8. Principles of Criminal Law
9. Roman Law I and II

To this already wide spectrum of required classes are a rich catalogue of electives, including classes in French and Italian. This is formally different from the 'trans-systemic' approach to legal education taken by, for example, McGill University in Quebec's mixed or poly-jural jurisdiction. There, multiple legal—and ideally other normative—orders are systematically taught within the same class. The Maltese approach is different, but nevertheless extends, on the whole, not only beyond state legal traditions, but into legal anthropology as well.[8]

Indeed, given the dominance of the domestic legal profession in legal education and the way in which the jurisdiction straddles and combines various legal traditions, it is not surprising that local educators do not perceive legal education as the wholesale adoption of a particular pedagogical model rooted in a homogenous understanding of law characterising a single dominant (external) legal tradition. Instead, they understand Maltese legal education as the elaboration of an indigenous model through selective borrowing from various sources, feeling empowered to draw selectively on foreign models.[9] Developments at EU level (including the Bologna process of harmonising, to a degree, legal education across Europe), in Anglo-American and Continental law, and Maltese statutes and jurisprudence are mined for inspiration. EU membership as a micro-state, the absence of a single dominant external power or 'Big Brother' on whom to model legal education, and the high degree of hybridity already present in the tradition fuse to promote a sense of empowerment in Maltese lawyer-lecturers and legal academics. They do not oppose the importation of foreign models, but arguably see themselves as actors of legal globalisation rather than its victims.[10] Apart from the revised, sui generis structure for the core law degrees which, this can also be illustrated by the eclectic use, referred to above, of English, Maltese and Italian texts in Maltese law-teaching.

Law students train as general practitioners or advocates. The Chamber of Advocates represents lawyers as a whole, particularly in relations with the government and regulatory bodies. There is neither an English-style distinction between

---

[8]See Donlan (2015) for a discussion of 'trans-systemic', 'supra-systemic', and even 'trans-temporal' legal education.

[9]Since 1993, over 2000 students have graduated with an LLD degree, making a total of more than 3000 individuals in possession of this degree relative to a population of 380,000.

[10]The Maltese language resulted from the bending of Semitic and Romance elements drawn from Arabic and Italian. Similarly, Maltese identity is asserted by fusing Semitic ethnic origins with European cultural influences embodied in English, the Catholic religion and Malta's defence of Europe from the Turks in early modern times and Nazi aggression in the Second World War.

barristers and solicitors nor an option for continental-style training for a judicial career. Given the difficulties of acquiring knowledge of both Anglo-American law and Continental legal rules and languages, as well as the use of Maltese in practice, outsiders find it almost impossible to meet the requirements for practice in Malta. Students may also opt to study to become notaries who, on the continental model, remain indispensable in legal practice, particularly as regards conveyancing, or even as legal procurators, who assist advocates and have some rights of audience in the lower Courts. Continuing education consists of a range of Masters degrees offered by the University, including Masters degrees in European and Comparative Law, European Business Law, Financial Services Law and Maritime Law. There is also the possibility of reading for a Masters or PhD degree in law by research. In order to specialise in certain fields which are not taught domestically, and to carve out an occupational niche for themselves in Malta's competitive professional environment, Maltese law graduates often go abroad for further education, with the UK and Italy being the preferred first and second choices to study for a Masters degree.

As in Anglo-American jurisdictions, Maltese judges are drawn from senior practitioners (rather than following the continental division, educational and professional, between magistrate and lawyer). Indeed, the Maltese Constitution requires 'Judges' to have at least 12 years experience in legal practice. By contrast 'Magistrates' are drawn from advocates who must have at least 7 years of experience of legal practice. The Courts, too, are more centralised than in most continental jurisdictions. Except for the division between civil and criminal matters, there are no specialised separate Court hierarchies. And unlike many jurisdictions, both Anglo-American and continental, there is no single superior Court over the ordinary civil and criminal Courts. There are, however, some specialised Courts within the ordinary Court structures. There are no, and have never been, Anglo-American Courts of Equity.

Unlike some other Commonwealth jurisdictions, Malta maintains neither an appellate link to Britain nor any institutional use of British or foreign judges. The possibility of appeals to the Judicial Committee of the Privy Council (hereinafter 'Privy Council') was removed when Malta became a republic in 1974; the main possibility of scrutiny by a foreign tribunal of decisions of the Maltese Courts is via the ECtHR and the European Court of Justice. In addition to some original jurisdiction related to Malta's House of Representatives, the Constitutional Court's appellate jurisdiction is limited to appeals from its Civil Court, First Hall. This includes matters regulated by the Constitution or Malta's European Convention Act, which incorporated the European Convention on Human Rights into Maltese law. Much more could be said about the Court, but it falls beyond the scope of this chapter.[11] And some questions remain about the Constitutional Court's role vis-a-

---

[11] See Donlan et al. (2012) for additional detail.

vis the Parliament; that is, whether the Constitution or Parliament is supreme (in the British manner).[12]

Given this context, globalisation has impacted on Maltese legislation and legal practice in ways which have been rather indirect and which have been buffered by choices made by local stakeholders. An example is the development of financial services. This required the enactment of detailed legislation and the establishment of the Malta Financial Services Authority in 2002. That sector has seen rapid growth over the past decade and a half. Initially financial services legislation was structured to cater for offshore registered companies. Later, however, the sector was unified as it was opened up to locally registered companies as well. In 2015/2016, the World Economic Forum's Global Competitiveness Index placed Malta amongst the top 20 financial jurisdictions in the world. Yet far from seeing this development as a threat to its legal tradition, most Maltese legal practitioners see the growth of the financial services sector as a process which has been guided by far-sighted reforms introduced by its legislators and which were themselves made possible by its mixed legal system.[13] This tendency to see financial services legislation as a natural development of the legal tradition rather than a foreign imposition can be seen in the way in which key elements of company and trust law have been integrated into the law curriculum alongside traditional civil law courses in Roman law, property law and the law of obligations. Thus students are taught to perceive principles of common law applicable to complex financial transactions as part and parcel of a broader set of Civilian legal rules and principles based on the Roman law. It can also be evidenced by the decision taken by the Law Faculty to offer a specialised Masters in Financial Services to be taken by students only *after* completing the LLD degree.[14] The degree complements rather than replaces the core law curriculum.

A similar approach can be seen in maritime law, another field which has grown substantially in recent years. In 2015, Malta's ship register was the largest in Europe and the sixth largest in the world. Here, too, the Government has built upon the early intervention of Arvid Pardo, Malta's ambassador to the United Nations (UN) in the 1960s, which eventually led to the drafting of the UN Convention on the Law of the Sea, to persuade the International Maritime Organisation (IMO) to establish a school at the University of Malta to offer an LLM in Maritime Law to law students from around the world. Directed by a Maltese legal academic, the

---

[12]This was debated in a series of articles and letters in *The Times of Malta* from April–June 2012. Giovanni Bonnello, former Judge of the ECtHR (1998–2010), argued that the Constitution was supreme (27 April, 2 May, 6 June, 20 June); Giuseppe Mifsud Bonici, Former Chief Justice and President of the Constitutional Court (1990–1995), argued that Parliament was supreme (2 June, 15 June). See also, Aquilina (2014) who more recently sides with Judge Bonello.

[13]Malta Financial Services Authority (2008), p. 4: 'It may be rightfully claimed that the Maltese legal system has been able to absorb a number of different legal and cultural influences from the two European countries which have shaped a large part of its history, and whose cultural influence remains very high to this day: Italy its closest neighbour to the north, and the United Kingdom'.

[14]The introduction of a new Masters in European Business Law also follows the same pattern.

school has since its inception offered two scholarships a year to Maltese law graduates. A similar story could also be told about e-gaming, another area of strong economic growth in Malta. On the whole, as in other areas of the legal system, globalisation is seen by the Maltese legal profession and domestic legal educators as reinforcing the Maltese tradition, rather than weakening it.

9.3 Jersey

The Bailiwick of Jersey is smaller still than the Republic of Malta, both in terms of population (100,000) and territory (118 km^2). Indeed, half of those on the island are not native to it. Nearly a third are from Britain or Ireland; 7% are from Portugal or Madeira. Jersey shares related histories and legal histories to the other Channel Islands, especially the Bailiwick of Guernsey, which includes Alderney, Sark, and other smaller islands. Each of these have broadly common histories and legal traditions, but differ in some respects and are largely governed by separate local institutions. Jersey is only 22 km from France and its legal history is closely related to continental developments, especially to Normandy. Part of the Duchy of Normandy, Jersey was held by the English Crown from 1066 and remained attached to it after the loss of Normandy. Jersey is neither part of the UK nor the EU. It is a 'Crown Dependency' of the former and has a special relationship to the latter, being treated as a member of the European Economic Community for the purposes of the free movement of goods. It is, however, self-governing, with its own Parliament (the 'States of Jersey') and ordinary Courts. There is a newfound sense of identity in recent decades; there has even been discussion of possible independence from the UK.[15]

As in Malta, language is an important aspect of the history of Jersey. The local language has never been modern standard French. Instead, it is Jèrriais, a variant of Norman. But both English and French are official languages in Jersey. English and Legal English began to displace both Jèrriais and French in the nineteenth century. This accelerated in the twentieth century, especially after the Second World War. As late as 2006, the conveyancing of immovable property was handled in French; since then it must be in English. Today, the languages of daily life and the law has shifted decisively to English. English has also been permitted in the States for over a century; French has little more than a ceremonial role there. As the knowledge of some French is still necessary for understanding its law, many practitioners have to have some training in French to pass their professional examination.

French remains relevant in a number of ways. A Jersey legal practitioner requires French, for example, for professional oaths. More importantly, French is needed to read many of its foundational texts of customary law and the commentary, from the sixteenth to the twentieth century, on those texts. It is required, too,

---

[15]See, for example, Bailhache (2010).

for most legislation before the 1940s, including some important matters, as well as amendment of that legislation. Almost all judgements of the Royal Court before 1950, as well as some into the 1960s, are in French, as is the *Tables des Décisions* (1885–1963), indices to unreported Royal Court judgements for that this period. And given the importance of French law and doctrine to Jersey, the language is also useful. But, as discussed below, a significant number of practitioners are not native to the jurisdiction and less likely than locals to read the language.

In addition to its local legal customs, Jersey law has drawn heavily on Norman customary law, modern French doctrine and law (including the *Code Civil* (1804)), and modern English legislation and jurisprudence.[16] As is the pattern in other mixed systems that experienced a superimposition of Anglo-American law on a continental foundation, English influence is especially strong in public and criminal law. Within private law, English influence is greatest in the area of torts. English-style trust law has also become important in the twentieth century. In contrast, French law has had a significant impact on Jersey's contract law. Not surprisingly, property law and succession law are the most resistant to English influence. Both reflect their Norman roots. Procedural law more clearly reflects an Anglo-British inheritance. As in Malta, there is no strict application of the doctrine of binding precedent or stare decisis within Jersey law. This was unclear until 1999. In *The State of Qatar*, the Royal Court rejected stare decisis by noting the Norman customary source of its law, the absence of the case law required for that doctrine, and the similarity of the Jersey tradition to the French tradition, in which binding precedent does not operate.[17] There remains, however, a strong presumption in favour of previous decisions, akin to the doctrine of jurisprudence constante of France and other continental jurisdictions. The Courts of Jersey are, however, bound by relevant decisions of the Privy Council. While Norman customary law remains important in the areas of property and succession, the modern sources of Jersey law reflect a strong English influence.

Unlike Maltese law and much of the law of Seychelles, Jersey law is largely uncodified, despite the influence of the *Code Civil* and the doctrine that informed it. And unlike the pull felt in Malta between the UK and Italy, there is little loyalty to continental law beyond those traditions already long-established in Jersey. A strong English influence is reflected in legislation enacted by the States. Since the 1930s, bills are drafted by individuals accustomed to the Anglo-American style of legislation where statutes are highly-detailed and precise. Legislated 'Laws' follow this style and are normally promulgated in English. In addition, Jersey legislation must be submitted to London for review by the Ministry of Justice and Royal Assent. The jurisprudence of the Jersey or Guernsey Court of Appeal and the UK Supreme Court is also important. Judgements of the Royal Court have, from the 1950s, adopted both the English language and the Anglo-American style. This is facilitated by the publication of the Jersey Law Reports since the 1980s. Doctrine

---

[16]This section draws on numerous sources, not least Bailhache (2014).

[17]*State of Qatar v Al Thani* (1999) JLR 118.

remains important. Indeed, its importance has arguably grown with the establishment of the *Jersey and Guernsey Law Review* in 1997 and the Institute of Law in 2008. And, a wealth of materials has been made available over the last decade online through the Jersey Legal Information Board.[18]

Jersey's approach to legal education and training reflects these influences, as well as the fact that the jurisdiction has never had a local university or formal Faculty of Law. In the nineteenth century, Jersey law students studied in France. Increasingly in the twentieth century, students typically studied at British, especially English, universities or other Anglo-American jurisdictions. In fact, before 1997, becoming an advocate or a solicitor in Jersey was possible with British, French or Irish qualifications. In practice, French study through the University of Caen fell into disuse. Legal education and training changed significantly with the creation, in 2008, of the Jersey Institute of Law. The Institute's mission was to be the central nexus for education and training, both academic and professional, in Jersey law. The Institute, housed in St Helier, Jersey's capital and largest urban area, does this in numerous ways. In addition to the facilitation and provision of legal education and training, it also works to coordinate research in selected areas: the Channel Islands, on mixed systems more generally, on micro-jurisdictions, and on offshore finance. It hosts events geared to legal academics, to legal practitioners, including Continuing Professional Development (CPD), and to the general public (in its 'Law Made Simple' talks). The Institute also runs the island's Law Library and works closely with the Law Society of Jersey, the Jersey Legal Information Board, and the *Jersey and Guernsey Law Review*.

The Institute's approach to legal education and training differs in significant respects from that of Malta. It started as an education institution focusing on the ab initio training of local lawyers in Jersey law. It then became evident, because of the regulations and legislations on the qualifications required to become such a lawyer, that the Institute could also provide a degree in English law. Without a university in Jersey, the Institute uses an academic, undergraduate LLB developed out of the University of London's well-known International Programme. That programme is taught globally. But its content is deeply rooted in the English legal traditions. As its 'Foundations of Legal Knowledge' component, the foundation of England's Common Law, the programme includes topics required by English regulatory bodies:[19]

---

[18]More information about the JLIB is available at https://www.jerseylaw.je (last accessed 7 July 2016). It operates as a 'repository of all laws and judgments for the Island of Jersey. It is [also] a site presented to the local and related professions as well as to the public generally so as to maximise access to legal information and services'.

[19]These are listed by the Solicitor Regulation Authority (SRA) in a Joint statement issued by the Law Society and the General Council of the Bar on the completion of the initial or academic stage of training by obtaining an undergraduate degree. Available at http://www.sra.org.uk/students/academic-stage-joint-statement-bsb-law-society.page. Accessed 7 July 2016.

1. Common law reasoning and institutions
2. Public law
3. Elements of the law of contract
4. Criminal law
5. Law of tort
6. Land law
7. Law of trusts
8. EU law

The programme also includes a class on Jurisprudence and Legal Theory, a dissertation (10,000 words), some optional subjects (Commercial law, the International Protection of Human Rights, etc.), and a 'Law Skills Portfolio' of legal research, oral communication, team working, information technology, and autonomy. Obviously these classes follow the English model, as if the graduate were to practice in England or Wales rather than in their native system. While most have some relevance in Jersey, others—particularly land law—are quite different. More generally, the result of these modules is to habituate students to Anglo-British conceptualisations about legal reasoning, the sources of law, and so on. While a more 'trans-systemic' approach is a possible option, other practical factors mean that that route has not been taken up to now.

Jersey's proximity to Britain means that the programme's faculty are largely drawn from there, even if teaching occurs in Jersey. Unlike ordinary university instruction of a few hours instruction during the week over the course of months, the classes are taught in longer, alternating blocks on weekends. This is necessitated by the number of non-resident 'flying faculty' and that students, largely drawn from the Channel Islands, are often employed during their studies, creating a very rich inter-connection between future employers and future lawyers. Given the small size of the student body—approximately 45 students a year for 2015–2016—teaching employs a mix of methods rather than pure lectures or tutorials, making the Institute one of the most successful registered centres of the entire international programme. The total number of student studying through the Institute is 180, which is quite large for a small island where approximately 450 students go to University annually (mainly in England). The Institute is now the largest provider of legal education on the island. The materials or study guides it produces are largely self-sufficient. Students with undergraduate degrees in other subjects have the option of an abbreviated Graduate Entry LLB. A Certificate in Higher Education in Law, also available through the University of London, may also be completed. It, too, focuses on English Law (same topics as the first year of the LLB). The Institute has also introduced a double degree in French and English laws. Students will be able to study both French and English law together. This may contribute to fostering a new kind of Jersey lawyer. Additional options for academic post-graduate study are available through the Institute's links with French and Spanish institutions.

But the Institute doesn't merely stand in for the academic component of legal education. It is also the hub for professional training in Jersey. Enrolment in its Jersey Law Course (JLC) is required for admission to the legal profession. This takes between 1 and 3 years. Approximately 150 students have completed the course. As with the Institute's academic teaching, the JLC is taught by visiting academics and local practitioners in intensive, seminar-style study weekends. The JLC syllabus includes the following:

1. Jersey legal system and constitutional law
2. Civil and criminal procedure, including professional ethics
3. Contract law
4. Law of security interests and bankruptcy
5. Law of immoveable property
6. Testate and intestate succession
7. Company law, Family law, or Trusts law

Tuition in beginners French and French for the study of contract law is also included. The JLC leads to the Jersey Law Examinations and 2 or 3 years of work experience with a Jersey law firm. Using the same classes, the Institute established an LLM in Jersey Law in 2015. Unlike the use of the London Programme, which inevitably orients students towards Anglo-British modes of thought, management of the JLC by the Institute has the potential to significantly increase the number of lawyers conversant with Jersey's legal traditions. Much depends, of course, on the relationship between the Institute and the Bar and the number of foreign-trained lawyers who undertake the JLC.

Jersey has a bifurcated legal profession consisting of advocates and solicitors (ecrivains). This division may appear similar to that between English barristers (as well as some continental legal systems). But Jersey advocates have direct access to the public, form partnerships, and handle client money. In addition, while they may specialise in Court advocacy and the preparation necessary for such advocacy, many will never appear in Court, but focus on transactional work related to international finance. Solicitors provide more general legal advice and assistance to individuals and companies. Founded in 1899, the Law Society of Jersey governs both advocates and solicitors. They promote the rule of law, the high standards and interests of its members, and the study of law. Notaries exist, too, though they are less significant than in the Maltese, Seychellois, or continental traditions. And, as elsewhere, Jersey's lawyers are required to engage in CPD.

As in Britain—as well as in Malta and Seychelles—judges are drawn from legal practitioners. There is no separate, continental-style entry into the judiciary. Among other functions, the Bailiff is the head of the judiciary, acting as chief justice, as well as President of the States. Both the Bailiff and the Deputy Bailiff are Crown appointments. In addition, Jersey, like the other Channel Islands, also has 'jurats'. Rooted in traditional positions with counterparts in ancien régime France, jurats once had a legislative role. Today they form, along with the Bailiff, Jersey's Royal Court. They are lay people who act to decide questions of fact, decide damages in civil matters, impose sentences in criminal matters, and preside over land conveyances and liquor licensing. Twelve in number, they are indirectly elected by an electoral college of members of the States and the legal profession. They serve until retirement at the age of 72. Other legal offices, including the Attorney General and

Solicitor General, exist as well. The former gives legal advice to the government, as well as acting at the head of Jersey's prosecution service; the latter deputises for the former.

Unlike Malta, Jersey maintains an appellate link to Britain and employs British judges from beyond Jersey. Before 1949, appeals could move from one chamber of the Royal Court to another or to the Privy Council. In that year, a Channel Islands Court of Appeal was established, but never sat. In 1961, a Jersey Court of Appeal was established. It includes the Bailiff and Deputy Bailiff, as President and Vice-President respectively, along with other judges. These judges are often English, though Scottish judges have also played an important role. This link to expatriate judges is not generally seen as negative in Jersey. Indeed, judges of the Court of Appeal appear to have been attentive to Jersey's legal tradition, notwithstanding the fact that they are not trained in that tradition.

Economic globalisation is a significant influence on Jersey's economy, laws, and culture. The island's role as a centre of international finance and legal services, and arguably as a 'tax haven', has altered significantly in the last few decades. Economically, globalisation may be seen to strengthen the standard of living of all of those in Jersey—half native, half newcomers—the influence on its laws and culture is more mixed. In the last half-century, the local Bar has changed from a small, inward-looking body to a large group that includes many lawyers, almost entirely British Anglophones trained exclusively in Anglo-British law, from beyond the island. These individuals have few links or loyalties to the local tradition. They focus almost exclusively on commercial law, including trusts. And because engagement with such areas doesn't require much new professional knowledge and is largely conducted through English and English-style procedures, the local Bar has found it difficult to manage outside actors and influences. The gravitational pull of British culture and English laws have had a significant impact on the cultural and legal identity of Jersey. It remains to be seen whether local developments, most notably the creation of the Institute of Law, might stabilise the state of the law or even reverse the effect of outside influence.

9.4 Seychelles

With respect to population, the Republic of Seychelles is the smallest of the jurisdictions considered here.[20] With some 92,000 people, it is slightly less populated than Jersey and less than a quarter of the population of Malta. It is, however, the most extensive, a relatively isolated archipelago of 115 islands with a combined territory of some 459 km². It is also the only jurisdiction of the three that is outside of Europe. Indeed, set in the Indian Ocean, it is some 1500 km east of Africa. This

---

[20]For more information, see Twomey (2014). See also Twomey (2015); Donlan and Twomey (2017).

distance meant that Seychelles was uninhabited by humans, at least in any permanent settlement, for most of its history. The French eventually settled the islands by way of Mauritius (the Isle de France until 1810). European settlers, along with their African slaves and indentured Indians, made various settlements from 1770 onwards. With the French Revolution and the subsequent Napoleonic wars, the archipelago became a pawn of imperial expansion. The French eventually lost the islands to the British in 1810.

As a result, the laws of Seychelles were for decades a mix of the plural French colonial *ius commune*, including the *Code Noir* for slave affairs. In the aftermath of the French Revolution, its laws altered in line with the changes being experienced in Europe, i.e. legal centralism, nationalism, and positivism. The *Code Civil*, which established a common national law for France was in effect in Seychelles by 1810. The same was true of the metropolitan commercial, penal, and procedural codes. With British rule, an Anglo-French legal hybrid was born. British legal influence slowly increased. English-style judgements were adopted from the middle of the nineteenth century. The English language would also become more widely spoken, especially in the decades after 1903, when Seychelles became a British colony in its own right. The British influence on criminal law was particularly strong. An English-influenced *Penal Code* was promulgated, for example, in 1955.

By the early twentieth century, a hybridisation or creolisation of cultures—European, African and Asia—had resulted. While the experiences of both Seychelles and Mauritius were similar in many ways, their modern traditions are distinctive. Perhaps most importantly, the French language is more commonly spoken there and, as a result, more important to its legal tradition. Seychelles is multilingual, too. English and French are both spoken, but a Seychellois creole (Kreol) is the day-to-day language of the great majority of the country. Since 1981, it is also an official language. But the language of legal practice is largely English. The result is a gap between present legal language and both (i) the language of the people and (ii) the original language of the law. And while the native legal profession can bridge the former gap, fewer and fewer are able to work comfortably in French. This is more complicated by the fact that non-native or expatriate judges may require linguistic translations. When those judges have been trained in a different legal tradition, related terminological and conceptual translations may also be necessary for them to perform their functions. And Seychelles' private law remains largely rooted in the French tradition, the law is now largely expressed in English. Perhaps most importantly, the Seychellois *Civil Code* was translated into English in 1975. In terms of precedent, too, the vertical binding effect of judgements are adhered to both in criminal law and civil law in practice despite the fact that the *Civil Code* provides that judicial decisions shall not be absolutely binding in civil matters.[21]

Academic legal study for the Seychelles Bar traditionally occurred beyond the jurisdiction. Before 1996, those admitted to the Bar of England and Wales and who had obtained a law degree were entitled to practise in Seychelles. This process

---

[21]Civil Code of Seychelles Act 1976, art 5.

changed significantly when the University of Seychelles (UniSey) was established in 2010. Alongside that development was the establishment of an LLB programme managed by UniSey's new Faculty of Law. As with Jersey, this is rooted to the University of London's International Programme, but is targeted to local students. Teaching is in English and the curriculum is rooted in the English legal tradition. But English contract and tort law, land law, and equity and trusts, are all inapplicable in Seychelles. And unlike the use of that programme by Jersey's Institute of Law, the great distance between Britain and Seychelles means that far more local instruction occurs, at least for now. Until 2015, this was provided by local practitioners, especially the local Bar. As part of the expansion of the Faculty, however, there are now three non-native faculty members who will teach the London modules. Each is European, though they represent both the Anglo-American and continental traditions. They are, unlike Jersey's 'flying faculty', residents of Seychelles. For the moment, the jurisdiction lacks the equivalent of the Jersey Legal Information Board. A similar site committed to preserving texts relevant to practice in Seychelles could be a very valuable resource.

Recognising the substantive distance between the London curriculum and Seychellois practice, additional instruction has also been introduced. The Bar Vocational Course (BVC) is broadly similar to the JLC and attempts to prepare participants to take the Bar Examination. It largely focuses on Seychellois law, including:

1. Civil Code
2. Code of Civil Procedure
3. Property Law
4. Penal Code and Criminal Procedure
5. General Principles of the Law of Evidence
6. Commercial Code [and] Company Law
7. Family Law
8. Constitutional and Administrative Law

A vocational module also focuses on drafting and conveyancing, Alternative Dispute Resolution (ADR), ethics (including bookkeeping and anti-money laundering (AML)), and advocacy. Unlike the JLC, the BVC is concentrated in a single-year programme taught in extended, three-hour sessions. Of course, this still occurs *after* immersion in the London programme and, as a result, habituation to Anglo-American legal concepts and legal thinking. Even more than in Jersey, a more 'trans-systemic' approach would seem to be a more appropriate choice. While much is still to being worked out, the new non-native, but resident faculty already chosen might, in practice, be ideal for such an approach, even within the limits of the London Programme. In any event, if a partner university is desirable, more appropriate choices may be possible. The choice of McGill University, for example, would be no less respected than the University of London. And the Quebecois and Seychellois legal traditions would be broadly similar.

It also remains possible in Seychelles to qualify for practice with an appropriate UK law degree, a degree from a Commonwealth country, or a French university, followed by successfully completing the BVC or the UK's Legal Practice Course.[22] Similarly, individuals may practice with a law degree from the University of Mauritius, followed by successful completion of the Mauritian Bar Exams or completion of the education phase of qualifying as a lawyer in France, followed by successful completion of the Seychelles Bar Exams. In each of these situations, but especially where students train in Anglophone jurisdictions, there is the possibility that Anglo-American legal habits and understanding will influence the local law. In addition, successful completion of the Bar Exam requires a pupillage (apprenticeship) of 2 years in an approved chamber.[23]

Lawyers are designated attorneys-at-law. They are generalists; there are no distinct categories of legal practitioners as operated during French or British rule. There are, that is, no distinctions between lawyers generally and those that can act as advocates before the Courts, as was the case both in the French tradition (avoués and avocats) and remains significant within English law (the division between solicitors and barristers).[24] The Bar Association of Seychelles represents and promotes its members and fosters the diffusion of legal information. It was only established in 1988. Most lawyers operate as sole practitioners, though there a few law firms as well. In addition, as in Malta, notaries remain important for drawing up authentic documents and performing conveyances. And, as in most jurisdictions, a CPD programme also exists.

With respect to the judiciary, judges are largely drawn from practice. Of greatest significance is the fact that, in addition to the selection of local lawyers, many of the judges are drawn from the Bench of other Commonwealth countries: e.g. the UK, Mauritius, Botswana, Dominica, Ghana, Hong Kong, Kenya, Lesotho, Nigeria, St Lucia, South Africa, Sri Lanka, Swaziland, Tanzania, Trinidad and Tobago, and Uganda. This practice has a long history, both in continental Europe and in Britain's colonies. Such judges arguably have a broader perspective, greater objectivity, or special experience that local jurists in a small jurisdiction might not have. Jersey and a number of Commonwealth jurisdictions still utilise the UK Supreme Court for appellate matters.[25] Even so, this need not be a positive experience. The use of expatriate judges may reasonably be objected to as a loss of legal autonomy. And

---

[22]See Legal Practitioners Act 1996, s 5(1)(a)(iv) which provides that these are degrees at institutions and of a level approved by the Minister in consultation with the Chief Justice.

[23]It is also possible to qualify after acting as an articled clerk over a long period of local study, consisting initially of an Articleship Entrance Exam, followed by 6 years of study at an approved chambers and successful completion of the local Bar exams.

[24]Distinct professions continue to operate in Mauritius. In 2011, the French merged representation by the avoués with the avocats before the Courts of Appeal. Rights of audience before the Cour de Cassation are still restricted to specialist advocates who are accredited to argue cases before the Cour de Cassation and the Conseil d'Etat, referred to as avocats au Conseil d'Etat et à la Cour de Cassation.

[25]Some of the jurisdiction of the Privy Council transferred to the UK Supreme Court in 2009.

the negative effects of non-native jurists may be compounded where the judge is from a tradition unrelated or distantly related to the local jurisdiction. This is especially true where a judge does not sufficiently comprehend the source language of the law to be interpreted and applied. Like Malta and Jersey, Seychelles is both a micro-jurisdiction and a mixed legal tradition. In such a situation, there may be a special danger of the imposition, conscious or unintentional, of imported legal norms, especially from the Anglo-American traditions. Again, a system of conti-nuous legal education exists for judges in Seychelles. This might provide a remedy to this problem.

Seychellois Courts include, in order of seniority, the Court of Appeal, the Supreme Court, the Magistrates' Courts, and other minor tribunals. The Court of Appeal is its highest Court of law and the final Court of appeal. The Court consists of a President, two or more Justices of Appeal and other judges who are ex-officio members of the Court. Until 2004, the Court of Appeal was not resident in Sey-chelles. The first Seychellois appellate judge was only appointed in 2005; its first Seychellois President in 2007. As a Court of last resort it has the opportunity and duty to correct mistakes and resolve conflicting decisions made by judges in previous cases. The hybrid nature of the law of Seychellois has, however, created tensions. The Supreme Court is the highest trial Court and the first Court of appeal from lower Courts and tribunals. The Supreme Court's jurisdiction is exercised in practice by a single judge. The Court can also sit as a Constitutional Court when handling public law matters. In this situation, at least two judges are required; normally three sit. Decisions of the Constitutional Court and the Supreme Court, in both civil and criminal matters either at first instance or on appeal, are subject to appeal to the Court of Appeal.

As with Malta and Jersey, globalisation has had a significant impact on the economy, laws, and culture of Seychelles. While this is unavoidable economically, it is also importantly a matter of law and legal traditions. The world economy and earlier Anglo-American legal influences have certainly affected Seychelles. In addition, however, the so-called new international rule of law dictated by organi-sations like the World Trade Organisation (WTO) and the International Monetary Fund (IMF) has also had an impact. These organisations often impose legal instru-ments foreign to the domestic legal system. In this way, policies rooted in Anglo-American doctrines and norms have proven to corrode the mixed nature of Sey-chellois law.[26] There are other more complex legal influences. Other international and regional organisations also have an impact on Seychellois law. These include the Commonwealth itself, as well as the Organisation Internationale de la Franco-phonie, the African Union, the Common Market for Eastern and Southern Africa (COMESA), and the Southern African Development Community (SADC). While Seychelles would admit into practice far fewer lawyers than Jersey, the fact that English—rather than Kreol—is used in practice means that there are more non-native attorneys than in Malta. With judges, however, Seychelles is far more

---

[26]See, for example, Mattei (2002).

open, arguably out of necessity given the small size of the Bar, to the influence, good or ill, of foreign judges. Finally, like Jersey, the creation of a new institution—here the Faculty of Law of UniSey—offers the possibility of generating lawyers more attuned, whatever their origin, to local needs.

9.5 Inside-Out/Outside-In

Micro-jurisdictions and small states engage with outside traditions and influences by necessity. Rather than making such systems marginal, this places them at the crossroads of interaction between peoples, jurisdictions, and powers. This is even more true of mixed legal systems. Those jurisdictions continue to draw on the complex, often colonial, legal traditions of other, parent states. The result of such legal, and often linguistic, borrowing is particularly obvious in micro/mixed jurisdictions like Malta, Jersey, and Seychelles. In each, legal ideas and institutions owe much to exterior actors and sources, especially perhaps those of the Anglo-British tradition. This reflects both their historical, colonial or quasi-colonial, links to Britain, as well as the continuing influence of Anglophone hegemony and the pan-national Anglo-American traditions. Such borrowing is particularly obvious in the legal education and training adopted and the organisation of the legal professions in those jurisdictions.

As indicated, Malta, Jersey, and Seychelles continue to orient themselves to other states, most notably Britain, but also to Italy and France. This is true of both substance and procedural law. But the professions themselves reflect their hybrid origins and the demands of their small jurisdictions. Perhaps as a practical matter, a bifurcated profession is less likely to arise in a micro-jurisdiction. In any case, both Maltese advocates and Seychellois attorneys-at-law are generalists. And if Jersey's advocates and solicitors superficially resemble British barristers and solicitors, it is important to note that such a division is not, or was not, unknown in France. Similarly, perhaps as a result of less pressure requiring the profound reform of legal orders that occurred elsewhere in the nineteenth century (centralism, positivism and so on), these extraordinary places may continue to maintain more archaic forms. And the relationship to Britain may take on many different shapes. This may involve a formal link to British appellate review, as in Jersey; this link is viewed positively. In Seychelles, while foreign Courts no longer judge local law, the judges themselves are often foreign. That this may be seen negatively may reflect the fact that those judges may be less committed to the local tradition than the senior jurists of the Privy Council in the past or the new Supreme Court in the present. These differences in perception may reflect the calibre of the judges chosen. They might also reflect linguistic obstacles, an issue that doesn't arise in Malta where fluent Maltese is required for practice at the Bar and on the Bench.

Of the three jurisdictions considered here, Malta is perhaps the least dependent on outside sources, in significant part because of a long-established, domestic third-level education in law at the University of Malta. Jersey and Seychelles, each smaller than Malta, have been historically more reliant on legal education abroad. As noted, language is another important factor. The Maltese legal profession operates without difficulty in English, both because the language is common and their legal education takes place in that language. But familiarity with the Maltese language is essential given its place in litigation and adjudication, as well as in the jurisprudence the Courts produce. The generation of this jurisprudence may insulate the system somewhat, as well as generating a body of judicial doctrine especially important in a jurisdiction in which scholarship is limited, at least in contrast to larger systems (merely by virtue of their size). And reference to legal materials and sources in Maltese—a language difficult for many outsiders to learn—no doubt produces some difficulty for visiting students and, more importantly, for non-nationals who might wish to practice there. Perhaps more importantly, Malta's size and long experience with third-level education means that its Faculty of Law is entirely composed of Maltese nationals. As these individuals are also usually practitioners, the result is that outside influence is filtered through native, practical legal experience.

In both Jersey and Seychelles, the absence of local legal education, at least at third level, historically required study abroad. This meant that such education typically included no study of their native system (though it may have been required locally and independently of instruction abroad). Instead, they were instructed in the British, or more precisely the English, tradition and were expected to apply, to varying degrees, the principles found there to their native system. And if the language of both daily life and the law has been English for some time in Jersey, the situation is more complicated in Seychelles. There, Kreol remains the language of the people, while the French language, and more importantly English, have been the language of the law.

Both Jersey and Seychelles have now brought legal education into their respective islands. But the law taught there, at least at the LLB level, remains a foreign law that, at least in the texts employed, has not been adapted for the local tradition. Indeed, the texts for both are the same products of the University of London. As Sue Farran has written in considering education of this type beyond England:[27]

> Although UK universities may not be consciously seeking to colonise legal minds and hearts…, there is the danger that a failure to consider the comparative strengths and weakness, or advantages and disadvantages of any transplant, or to conjointly develop appropriate degree course with non-UK counterparts, is seen as an attitude of empire.

And perhaps by necessity, these small micro-jurisdictions also rely on faculty that are overwhelmingly drawn from beyond their shores. While not necessary, many of these individuals have never practised, either in the jurisdiction or any other. As a result, there is the danger that legal education may be abstract and less clearly tailored to practice, as well as foreign or modified through a foreign lens.

Legal educators in Jersey and Seychelles have sought to limit the impact of using foreign materials by requiring additional study of local laws. In Jersey, this occurs

---

[27]Farran (2013a).

in the JLC, as well as in the requirements of legal internship. In Seychelles, an additional year of legal education will occur after the London LLB; this is followed by preparation for bar examinations and internship. While the University of London programme brings obvious benefits, not least in making local legal study attractive to those who want some exterior validation of their work, perhaps with the prospect of practising in England, Wales, and Northern Ireland. Indeed, in Jersey at least, there is the real possibility of attracting non-national Britons to the island for their legal study before returning to practice in England, Wales, and Northern Ireland. But a more comparative, even trans-systemic, legal education might also be considered and be of more value to those who ultimately practice in Jersey and Seychelles. Malta has taken a similar approach already, even extending its programme beyond state law to legal anthropology. And if foreign validation is desirable, and it's reasonable to argue that it is, there might be more appropriate institutions with which the Jersey's Institute of Law and UniSey's Faculty of Law could ally themselves.

Finally, jurisdictions like Malta, Jersey, and Seychelles might also have important lessons to offer on globalisation. They are not, of course, the only legal systems influenced by exterior or foreign global forces. No state, however large, is immune from such influence. The particularly small size of micro-jurisdictions, however, and the place of mixed systems at the intersection of different legal traditions may make systems that combine these elements especially fragile. But as Malta and, to a lesser extent, Jersey suggest, their size and situation might allow such jurisdictions to act nimbly, as well as the luxury, or possibility, of knowingly choosing options drawn from more than one legal tradition. Finally, there is the question of access to sources, not least through source languages that may not, or may no longer, be understood by a nation's lawyers or its public. More importantly, such a gap between past and present raises complex questions about practical and prescriptive fidelity to tradition and, indeed, of the reification of legal traditions themselves. Legal hybridity or mixity invariably touches on issues of space (i.e. where the law has come from?), but also of time (i.e. when was the law received or the legal link established?). They raise the question of why the people of the present are ever attached to laws generated by the past. Indeed, whatever the link between state and society, meaningful self-government is more important, allowing nations to remedy deficits in their laws. All legal traditions are hybrid mixes of autochthonous and imported elements.[28] Handled with care, openness to outside influence need not prevent effective choices about the substance or spirit of the laws. It may even empower those choosing.

---

[28] See, for example, Donlan (2014).

References

Andò B, Aquilina K et al (2012) Malta. In: Palmer VV (ed) Mixed jurisdictions worldwide: the third legal family, 2nd edn. Cambridge University Press, Cambridge

Aquilina K (2011) Rethinking Maltese legal hybridity: a chimeric illusion or a healthy grafted European law mixture? J Civil Law Stud 4(2):261–283

Aquilina K (2014) Do pronouncements of the constitutional court bind *erga omnes*? The common law doctrine of *stare decisis* versus the civil law doctrine of nonbinding case law within a Maltese legal context. In: Palmer VV et al (eds) Mixed legal systems, east and west. Ashgate Publishing, Farnham

Bailhache P (ed) (2010) Dependency or sovereignty? Time to take stock. Jersey & Guernsey Law Review. Published transcript of conference proceedings, Institute of Law, Jersey, September 2010

Bailhache P (2014) Jersey: avoiding the fate of the dodo. In: Örücü E et al (eds) A study of mixed legal systems: endangered, entrenched, or blended. Ashgate Publishing, Farnham

Bogdan M (1989) The law of Mauritius and Seychelles: a study of two small mixed legal systems. Juristförlaget i Lund

Castellucci I (2014) San Marino, Città del Vaticano, Malta: Il diritto italiano nelle micro-giurisdizioni. In: Annuario di diritto comparato e di studi legislative. Edizioni Scientifiche Italiane, Napoli

Donlan SP (2014) Things being various: normativity, legality, state legality. In: Adams M, Heirbaut D (eds) The method and culture of comparative law: essays in honour of Mark van Hoecke. Hart Publishing, Oxford

Donlan SP (2015) Everything old is new again: stateless law, the state of the law schools and comparative legal/normative history. In: Deldek H, Van Praagh S (eds) Stateless law: evolving boundaries of a discipline. Ashgate Publishing, Farnham

Donlan SP, Twomey M (2017) Island, intersection, or in-between? Legal hybridity and diffusion in the Seychellois legal tradition, c1750–1950. In: Donlan SP, Mair J (eds) Mixity, metaphor, and marriage: Essays in honor of Esin Örücü. Routledge, Oxford

Donlan SP, Andò B et al (2012) "A happy union"?: Malta's legal hybridity. Tulane Eur Civil Law Forum 27:165–208

Farran S (2013a) The "age of empire" again? Law Teacher 47(3):345–367

Farran S (2013b) The age of empire. again: critical thoughts on legal imperialism. Plenary address delivered to the Irish Society of Comparative Law Annual Conference, Galway, Ireland, 24 May 2013. Available at https://www.youtube.com/watch?v=UcHLfEofEpo&list=PLHoyytvdqQKz UF79lmo5xdrjBrb5e59yC. Accessed 9 July 2016

Malta Financial Services Authority (2008) Malta: a framework for financial services. Available at http://www.simonestates.com/business2.pdf. Accessed 9 July 2016

Mattei U (2002) A theory of imperial law: a Study on US hegemony and the Latin resistance. Indiana J Glob Leg Stud 10:383–448

Palmer VV (ed) (2012) Mixed jurisdictions worldwide: the third legal family, 2nd edn. Cambridge University Press, Cambridge

Twomey M (2014) Seychelles: things fall apart? – The mixing of fate, free will and imposition in the laws of seychelles. In: Örücü E et al (eds) A study of mixed legal systems: endangered, entrenched, or blended. Ashgate Publishing, Farnham

Twomey M (2015) Legal salmon: comparative law and its role in Africa. In: Mancuso S, Fombad CM (eds) Comparative law in Africa: methodologies and concepts. JUTA and Company, South Africa

Zammit EL (1984) A Colonial Inheritance: Maltese perceptions of work, power and class structure with reference to the labour movement. Malta University Press, Malta

Chapter 10
On Law, Legal Elites and the Legal Profession in a (Biggish) Small State: Cyprus

Nikitas E. Hatzimihail

10.1 Introduction

An island country with a population around the one-million mark, Cyprus would fancy itself a great power among small states as traditionally defined.[1] As a matter of fact, Cyprus competes with Iceland and Luxembourg for top place at the biennial Games of the Small States of Europe—and it should come as no surprise that we swept the 2009 games held in Cyprus.[2] As a biggish small state, Cyprus offers therefore an interesting case to study in elaborating a theory about the role of law, and lawyers, in small states (and island countries).[3]

---

[1]I am adhering to the traditional definition of small states as countries with a population of less than 1.5 m. See The World Bank (2016). For a short, albeit very enlightening and comprehensive, examination of possible criteria see Thorhallsson and Wivel (2006), pp. 652–655.

[2]The Games of the Small States of Europe (GSSE) were launched in the 1984 Olympics by the National Olympic Committees of Andorra, Cyprus, Iceland, Liechtenstein, Luxembourg, Malta, Monaco, San Marino, to which was added Montenegro in 2009. They are being held in odd years since 1985. In the 2009 games, Cyprus had its best performance, with a total of 139 medals, followed by 81 for Iceland, 62 for Luxembourg and 42 for Monaco. In the 2015 games, held in Iceland, the host country came first with 115 medals, followed by Luxembourg (80), Cyprus (52) and Monaco (33). There is clear correlation between the size of each country's *team* and its medal count.

[3]The monograph by Thorhallsson (2000) has been especially influential on the development of what is a growing body of literature regarding, on the one hand, small states in international relations, and, on the other hand, European small states and their interactions within/with the European Union. Thorhallsson has also done much to present Iceland as an interesting small state case study. See especially Thorhallsson (2004). Of more recent work, see the collection of essays in Baldersheim and Keating (2015).

N.E. Hatzimihail (✉)
Department of Law, University of Cyprus, Nicosia, Cyprus
e-mail: nhatzimi@ucy.ac.cy

© Springer International Publishing AG 2017
P. Butler, C. Morris (eds.), *Small States in a Legal World*,
The World of Small States 1, DOI 10.1007/978-3-319-39366-7_10

Cyprus has been a British colony until 1960, an EU Member State since 2004 and a Republic not in control of over a third of its territory since 1974 (and of almost a fifth of its population since 1964). Considered together, these factors should make it an interesting case to study regardless of its size. Much of the legislation and the legal institutions—and even many of the law buildings—still in place have a distinctively colonial and/or post-colonial flavour; moreover, the complicated political situation has led to a combination of a traditionalist mentality with the sense of *perpetual temporariness*. At the same time, Cypriots—certainly the ethnically dominant Greek Cypriots—have strongly identified with Europe and have been quite happy to partake of European law and participate in EU institutions, even though they are only gradually realising the scope of change ushered in by European integration.

This chapter has twin goals. On the one hand, to introduce Cyprus to the emerging international scholarship on small-state governance.[4] On the other hand, to explore how size is impacting the role—and functions—of law and lawyers in a small state.[5] This study draws on my ongoing research on the Cyprus legal system from a comparative-law point of view.[6] It is also undoubtedly influenced from my experience on the ground, as a law teacher with some experience in Cyprus practice and policy making.

My study may be best understood in the context of modern comparative-law scholarship, especially the work of the past two decades on mixed legal systems—both in the classic definition of legal systems in which common-law and continental legal tradition coexist in a sustainable fashion and the more expansive notions entertained in recent scholarship.[7] In the past few years, the term *hybridity* has also been used to account wholesale for what one may call phenomena of—individualised as well as sustainable—coexistence of different legal traditions within an individual country, and the attempt of comparative lawyers to move beyond the traditional taxonomies of the discipline.[8] Moreover, as will become apparent throughout the chapter, comparative-law concepts about legal change are especially important in my study of Cyprus law. The concept of *legal transplant* is an especially important tool for understanding the formation and evolution of a legal system such as Cyprus.[9] Much, indeed most, of Cyprus legislation has been

---

[4]The only other contribution I have located is Theofanous (2016), pp. 28–49.

[5]In that sense, my research (although at an early stage and with different aims and perspective) may be seen as building on the work of political scientists, such as Baldur Thorhallsson, on the role of political and bureaucratic elites in small states, including how their sensibilities impact on their state's orientation.

[6]See Hatzimihail (2013, 2015).

[7]The principal reference work is still Palmer (2012). Among the theoretical expositions see notably Palmer (2008) and Örücü (2008).

[8]See Castellucci (2012), pp. 668–672 and 696–720; Cassin Ritaine et al. (2010).

[9]The term is credited to Alan Watson. See Watson (1993). For a concise presentation of the concept and its critics see Fedtke (2006), pp. 434–437. For an intellectual history see Cairns (2014).

effectively imported from abroad. In many instances, such importation has resulted in veritable *transplants* in the literal sense of the word. In other instances, however, the purported transplants have found a different use within the Cyprus legal system or they have resulted in the formation of altered legal regimes. Although such alterations often evoke Teubner's concept of *legal irritants*,[10] for the purposes of this study, which is not intended as an active contribution to comparative-law theory, I have chosen the more generic term of *mutation* to account for these phenomena.

The first two sections of the chapter serve as an extended introduction to Cyprus law. The remainder of Sect. 10.1 presents some fundamental facts and observations. Section 10.2 presents the basic notions and institutions pertaining to the administration of justice (structure of the judicial system and organisation of the legal profession). The next two sections go further in examining the impact of being a small state on the law and the lawyers. Section 10.3 considers three areas of law that have developed in distinct directions. Section 10.4 elaborates on the allocation of power within the legal profession and identifies the most influential groups within the legal establishment.

10.1.1 Cyprus Situated

With an area of 9251 km^2, Cyprus is the third largest island in the Mediterranean, behind Sicily and Sardinia.[11] Officially, the island's total population falls just below the one-million mark.[12]

Since the Second Millennium BCE, the overwhelming majority of its population has identified as ethnic Greek. However, the island is geographically much closer to Anatolia (Turkey) to its north and the Middle East to its east than it is to the Greek mainland to its west. Its location has been a principal cause of both its strategic importance and its—past and present—misfortunes.[13] For the last 800 years, Cyprus has been ruled successively by the Lusignan Kings of Jerusalem, the Republic of Venice, the Ottoman Empire and the British Empire.[14] When the British took over the island in 1878, the legal environment mixed Byzantine

---

[10]Teubner (1998).

[11]Ministry of Foreign Affairs of the Republic of Cyprus (2012).

[12]In 1960, the year of independence but also of the last island-wide official census, the island's native population was estimated at 550,000 people, composed of 81.14% Greek and 18.86% Turkish Cypriots. See Solsten (1991). The ethnic proportion of roughly 4:1 is a sensitive point and still adhered to, but the latest census (December 2011) presents a population of 681,000 Greek Cypriots (including 8400 members of the Maronite, Armenian, and Latin groups), 90,100 Turkish Cypriots and 181,000 foreign residents. This count does not include the so-called 'settlers' from mainland Turkey (estimated by some at 160,000).

[13]See Mallinson (2008).

[14]Hill (2010) emphasises this period.

Roman law, as administered by the institutions of the Greek Orthodox Church, and Islamic law with Western-styled secular Ottoman legislation.[15] To this mix was added, at first, British procedural law. When Cyprus formally became a British colony in 1925, English substantive law began its conquest of the land. By the time of independence in 1960, Cyprus would be regarded as a small but definite member of the common law tradition.[16]

On the face of it, independence should have led Cyprus away from the common law tradition: after all, it supposedly empowered a people attached to mother-lands—and languages—falling firmly within the Continental legal tradition. But independence (imposed as it was on the population at large) effectively confirmed the place of the colonial legal, business and administrative elite, as well as its capacity to absorb new entrants. The institutional arrangements put in place in 1960 would have disallowed significant law reform anyway. Greek Cypriots, who had fought for union with Greece, were instead called to share power—in a manner viewed by most as disproportionate to ethnic and social demographics and even less favourable than colonial-era de facto quotas—with the ethnic Turkish minority (whose leadership flirted with the idea of the island's partition between Greece and Turkey). The Constitution established a presidential system of two separate *communities*, Greek and Turkish.[17] For each Greek Cypriot in a constitutionally pro-scribed government office, a Turkish Cypriot deputy with veto power was appointed.[18] The House of Representatives effectively consisted of two caucuses, with separate majorities needed for matters such as taxation, electoral reform and city government.[19] Relative parity was provided in the two supreme Courts of the land and quotas in the civil service. Separate municipal structures—both mayors and city councils—were elected in the five major cities.[20]

The political conditions soon deteriorated into a state of emergency in 1963–1964, culminating in a bombardment by Turkey, the departure of the Turkish Cypriot leaders from their government posts and the segregation of Turkish Cypriots into autonomous enclaves. The Republic continued to operate on the fiction that Turkish Cypriots would return to their positions, although amendments were made in the function of certain institutions under the so-called necessity

---

[15]Hatzimihail (2013), pp. 43–44.

[16]Ibid., pp. 40, 44–48.

[17]The Constitution of the Republic of Cyprus, art 2. The Constitution acknowledges three non-Greek Orthodox religious groups (Armenian, Maronite Catholic and Latin Catholic), all three of which elected in 1960 to join the Greek Community pursuant to Article 2(3).

[18]Ibid., art 1 (President and Vice-President of the Republic); art 72(1) (Speaker and Deputy Speaker of the House of Representatives); with regard to the *independent officers* of the Republic, art 112(1) (Attorney-General and Assistant Attorney-General), 115(1) (Auditor General and Assistant Auditor-General), 118(1) (Governor and Deputy Governor of the Central Bank), 126 (1) (Accountant-General and Deputy Accountant-General).

[19]Ibid., art 62 and art 78(2).

[20]Ibid., art 173.

doctrine.[21] The fiction has persisted even after the two-stage Turkish invasion in 1974 effectively divided the island. Today, the Republic of Cyprus controls a little less than two thirds of its actual territory, while the vast majority of Turkish Cypriot residents of the island effectively live in the northern territory (36%) controlled by Turkey and its military. The Republic's institutions retain their bicommunal orientation. Turkish Cypriots are citizens of the Republic and can travel freely in its territory and the rest of the European Union but, apart from those few who are resident in territory under the control of the Republic, they do not participate in its political institutions pending solution of the 'Cyprus problem'. Another 3% of the territory constitutes a buffer zone under the control of U.N. peacekeepers, while the United Kingdom claims sovereign status for its two military bases, Akrotiri and Dhekelia, which cover another 2.74% of the island.

10.1.2   In Comparative Law Terms

Comparative law scholarship has only very recently begun to take notice of Cyprus and to acknowledge it as a mixed jurisdiction.[22] But to understand Cyprus law, one has to begin by thinking in common law terms. In fact, complicated though the British colonial legal history of the island as it may be, by the time of independence in 1960, Cyprus would be regarded as a small but definite member of the common law legal family. To this day, the common law has maintained its hold on Cyprus law. Contrary to classic mixed jurisdictions, the heart of private law, namely the law of obligations, is a stronghold of English common law in Cyprus. Criminal law is another privileged domain of the common law.[23]

The unitary Court structure at the heart of the administration of justice system, with one trial and one appellate instance, certainly looks like a common law judiciary. In fact, the present judicial structure of Cyprus is principally a legacy of the late colonial period. This was ensured by the merger, a few years after independence, of the two supreme Courts originally provided for in the Constitution. The Cyprus judiciary strongly identifies with the common law tradition—an attitude shared by much, though by no means all, of the legal profession at large—and uses common law tools in judicial reasoning. Moreover, procedural law has been a vital tool for the continued dominance of the common law—even in areas where substantive law is modeled after, or even transplanted from, continental law.

However, pockets of resistance to the common law have been expanded or created. Let us consider, for example, family law: by the end of British colonial rule, only divorce was left to ecclesiastical jurisdiction; the family law reform in the early 1990s, however, not only transplanted secularised Greek family law across the

---

[21]Polyviou (2015) and Kombos (2015).

[22]Symeonides (2003) and Hatzimihail (2013, 2015).

[23]Papacharalambous (2015).

legal field but also extended its effective application to all Cyprus domiciliaries. But the principal example of resistance against the common law concerns public law: Greek administrative law (itself developed by case law in the spirit of the French administrative legal tradition) became the principal source of the new administrative law of Cyprus; today, the general principles of Cyprus administrative law are codified in statute and the case law of the Greek Council of State continues to be held in high esteem.

A closer look, however, at the operation of Cyprus Courts, the structure of the bar and especially of the judiciary, will demonstrate considerable elements of hybridity and mutation.

10.2   The Administration of Justice System

10.2.1   The Courts

Cyprus presently maintains a unitary—at least in principle—two-tier judicial system, one level each of trial and appellate jurisdiction. The appellate jurisdiction of the Judicial Committee of the Privy Council, which acted as a third-instance Court of last resort during the Colonial period, was abolished immediately upon independence.[24] The Constitution explicitly designates the High Court (subsequently renamed the Supreme Court of Cyprus) as the highest Court of last resort ('supreme second-instance court') and allows lower Courts to be established by statute.[25]

The judiciary is supported by an administrative mechanism of registrars and law clerks. The Court administrative personnel are considered part of the civil service of Cyprus: appointments and promotions are thus controlled by the Civil Service Commission. At the head of Court administration sits the Chief Registrar of the Supreme Court. The Supreme Court employs permanent law clerks ('Legal Officers'), who assist the Justices with research and in drafting their opinions, especially with regard to administrative law cases. Originally modeled after the law clerks of common law appellate Courts, these legal officers increasingly play a role similar to—and certainly identify themselves with—the Assistant Judges (εισηγητές; *Auditeurs* in French) of the Greek Council of State (who constitute, however, junior members of the judiciary, and tend to rise through the Court's ranks).[26] At present—and somewhat controversially—these, too, are considered as civil servants, rather than judicial officers.

---

[24]Cyprus Act 1960, c 52, s 5.

[25]The Constitution of the Republic of Cyprus, art 152(1) explicitly provides art 155(1).

[26]The *Référendaires* of the European Court of Justice are another model alluded to; however, the legal officers of the Supreme Court have a very high rate of permanent service. Very few have moved. A few have been subsequently appointed as Family Judges.

10.2.1.1   Justice at First Instance

The primary trial Court, i.e. the Court of general jurisdiction, is the District Court (Επαρχιακό Δικαστήριο).[27] Its jurisdiction extends over most civil and criminal matters.[28] The Supreme Court retained trial jurisdiction over admiralty cases,[29] but since 1986 the District Court is enabled to hear certain kinds of admiralty cases referred to it by the Supreme Court (whether on its own initiative or by application of a litigant).[30]

All cases are judged by a single judge—with the exception of serious crimes judged by the Assizes Court (Κακουργιοδικείο), which sits in panels of three rotating senior District Judges.[31] Specialised tribunals, consisting of one professional and two lay judges (one representative for each of the respective social groups), adjudicate rent-control cases and employment disputes.[32]

The trial Courts of Cyprus are staffed by professional judges with tenure. With the exception of the representatives of the social groups participating in the Employment and Rent Control Tribunals (and the representatives of religious groups), no lay participation is provided for anywhere in the administration of justice system. Moreover, there exist no magistrates' Courts—or other small-claims jurisdiction.

The Family Courts are one of the most interesting examples of Cyprus' legal hybridity. In 1990, as part of a comprehensive reform of family law, the Republic established secular Family Courts.[33] Until then, family law had been a matter of personal law, administered by community tribunals.[34] The British had removed community jurisdiction over a range of matters, including childcare and marital property, leaving ecclesiastical Courts with jurisdiction over the validity and dissolution of Greek Orthodox marriages.[35] Article 111 of the Constitution maintained the application of jurisdiction of Greek Orthodox ecclesiastical Courts.[36] The Family Courts were originally intended to replace community

---

[27]Courts of Justice Law 1960, art 22(1).

[28]Ibid. These cases are listed in an Annex to the Law.

[29]Ibid, art 19.

[30]Ibid, art 22B (added by L. 96/1986).

[31]Ibid, art 5. The Assizes Court is presided over by a Judge-President of the District Court, with two Senior District Judges (or District Judges) as members. The Law does not dictate the duration of the term. Members of the Assizes Court may also sit in regular District Court cases.

[32]On the Employment Tribunal, see the Remunerated Annual Leave Law 1967 (L. 8/67, as amended by L. 5/1973, art. 3), arts 12 and 12A; Termination of Employment Law 1967 (L. 577/67), arts 30–31. On the Rent Control Tribunal, see the Rent Control Law 1983 (L. 23/83), art 4. The Court Martial is usually regarded as a tribunal.

[33]Family Courts Law 1990 (L. 23/90), as amended.

[34]See Serghides (1988).

[35]See art 34 of the Courts of Justice Law 1953 (L. 40/53, Cap. 8). For English-era family legislation see Caps. 274–280 in the Statute Laws of Cyprus, vol. 5.

[36]See the original art 111 of the Constitution of the Republic of Cyprus.

tribunals, especially with regard to Greek Orthodox Cypriots. Separate Family Tribunals of Religious Groups were also set up to deal with the divorces of members of the three religious groups (Armenian, Maronite, Latin) recognised by the Constitution: these tribunals are composed of one 'President' judge, appointed by the Supreme Court 'from among members of the judicial service', one District Judge, and one representative of the respective group.[37] Over the past two decades, the jurisdiction of the Family Courts has been expanded, by statute and via the case law of the Supreme Court, both *ratione materiae* on every aspect of family law, and *ratione personae*.[38] Today, the Family Courts of Cyprus appear not unlike the Australian Family Court, or the Family Division of the High Court in London. They are viewed as Courts of specialised jurisdiction, not as tribunals.[39] At the moment, the main exceptions to Family Court jurisdiction concern the validity and dissolution of marriage under the rules of the three religious groups (which fall under the jurisdiction of the respective Family Tribunal), and some cases involving Turkish Cypriots.[40] A three-member panel of Supreme Court justices, rotating in two-year terms, sits on appeal as the 'Second-Instance Family Court'.[41]

10.2.1.2 Justice at the Appellate Level

At the apex of the administration of justice system sits a single appellate Court: the Supreme Court of Cyprus. The 13-strong Supreme Court has the attitude, and powers, of a common law Court of last resort. Its status and powers are determined in detail by the Constitution.[42] The Supreme Court sits on appeals and supervises trial Courts and tribunals.[43] The Justices of the Supreme Court[44] act as the Supreme Judicial Council, which selects, appoints, promotes, and moves trial judges around.[45] The Supreme Court also writes the Rules of Procedure, which govern

---

[37]See the Family Courts (Religious Groups) Law 1994 (L. 87(I)/94), especially art 3.

[38]See a full account of the evolution in Hatzimihail (2017, forthcoming).

[39]*Sioukrou v Ulrich* (2011) 1 C.L.R. 443 (Kramvis, J apparently endorsing a statement to that effect by Nikitas Hatzimihail in [2008] 1 *Lysias* 47).

[40]See Serghides (2010), pp. 24–29.

[41]Family Courts Law 1990, art 21(1).

[42]The Constitution of the Republic of Cyprus, arts 152–163.

[43]Ibid, art 155(1); Courts of Justice Act 1960, art 25.

[44]The same word (Δικαστής) exists in Greek for 'Judge' and 'Justice' (as a person's title). Given that the Constitution referred to the members of the High Court as Judges, reference to them in English as Justices has been a very recent phenomenon.

[45]Ibid, art. 153(8) and 133(8) respectively. Administration of Justice (Miscellaneous Provisions) Act 1964 (L. 33/64), art 10, as amended by L. 3/87, art. 2). Between 1964 and 1987, Supreme Court justices constituted only the plurality of the Council (which was composed by the Attorney General, the President and two justices of the Supreme Court, one President of a District Court, one District Judge, and one experienced advocate). The Constitution provided, in art 157, that the High Court Judges act as the Supreme Judicial Council. Supreme Constitutional Court judges were to act as judicial council for matters pertaining to the High Court Judges, and vice versa.

most procedural matters.[46] It issues prerogative writs.[47] It is also the country's constitutional Court, with full power of judicial review in full bench,[48] as well as—until early 2016—the sole administrative Court.

The Constitution had provided, in fact, for two supreme Courts: the *Supreme Constitutional Court*[49] and the *High Court of Justice*,[50] the former with one Greek and one Turkish Cypriot member, the latter with two Greek and one Turkish Cypriot members.[51] In both, a 'neutral judge' (i.e. not a national of Cyprus, Greece or Turkey) was to preside—casting the deciding vote in cases of disagreement.[52] Constitutional Court members were supposed to act as a supreme judicial council overseeing the High Court judges, and vice versa.[53] The High Court reprised the first- and second-instance jurisdiction of the colonial Supreme Court over civil and criminal cases.[54] The Constitutional Court was, in addition to what is implied in its name, endowed with trial-instance jurisdiction over administrative-law cases.[55] Unlike his High Court counterpart, the President of the Supreme Constitutional Court could not be a 'subject or citizen' not just of the United Kingdom but also of any of its present-day or former colonies.[56] The German law professor Ernst Horsthoff was accordingly appointed to that position, whereas the first High Court President (1960–1961) was the Irish Barra O'Briain, and his successor the Canadian John Leonard Wilson (1962–1964).[57] In 1964, as the constitutional crisis escalated and the threat of full-scale war loomed over Cyprus, both foreign Presidents left the island and the House of Representatives decided to merge the two high Courts into a single *Supreme Court of Cyprus*.[58] It must be noted that the term

---

[46]The Constitution of the Republic of Cyprus, art 163.

[47]Ibid, art 155.4. The writs include habeas corpus, *certiorari, prohibition, mandamus,* and *quo warranto*. See the overview of case law in Artemis (2004).

[48]*Board for Registration of Architects v. Kyriakides* (1966) 3 C.L.R. 640.

[49]The Constitution of the Republic of Cyprus, arts 113–151.

[50]The Constitution of the Republic of Cyprus, arts 152–164.

[51]Ibid, arts 133(1) and 153(1) respectively.

[52]Ibid, art 133(1)(1) for the Supreme Constitutional Court; Art 153 for the High Court (whose President was to have 'two votes', in order to balance off a two-judge plurality).

[53]Ibid, art 153(8) and art 133(8), respectively.

[54]Ibid, arts 155–156.

[55]Article 146 of the Constitution establishes a general ground of jurisdiction over petitions (the term 'recourses' is being used, in the spirit of the French *recours en annulation*) to annul or confirm administrative acts. The provision has acquired great importance in actual practice. The Constitution also empowered the Supreme Constitutional Court to make a final determination of cases where the Public Service Commission is unable to muster the necessary majorities for an appointments or promotion decision. See ibid., art 125(3), in conjunction with 151(1).

[56]Ibid, art 133(2)(3).

[57]Hadjihambis (2010). See pp. 112–13 on Forsthoff, pp. 114–15 on O'Briain, and pp. 120–21 on Wilson.

[58]Administration of Justice (Miscellaneous provisions) Law 1964 (L.33/64), especially art 3.

in Greek (Ανώτατο Δικαστήριο), is the same for both the High Court of Justice and Supreme Court. The two Turkish Cypriot incumbents continued to participate for a few more years, and indeed the Turkish Cypriot High Court Judge Mehmet Zekia (1903–1984) became the united Supreme Court's first President, on the basis of his seniority to the bench.[59] A side effect of the merger was that the High Court, which represented the continuation of the British colonial tradition and the English common law, effectively absorbed the Constitutional Court, which had embarked upon a process of transplantation and development of Continental public law doctrine. Even though the Continental doctrinal influence over Cyprus administrative litigation persisted (in fact, it was significantly expanded) since 1964, it is likely that it would have had a more systematic, less haphazard character were it emanating from a specialised appellate bench with a Continental orientation, as opposed to a Court with a strong, almost exclusively common law identity.[60]

Today, the Supreme Court constitutes a real super-Court, which has absorbed the powers of both Courts—and has more recently extended its jurisdiction over family law matters, previously left to confessional Courts. The Supreme Court constitutes the veritable final arbiter of constitutional questions, given that the Constitution may only be modified with a procedure based on the doctrine of necessity, in cases it is absolutely necessary to do so.[61]

It is therefore evident that the Supreme Court of Cyprus is not just your typical 'patriarchal' common law highest appellate Court.[62] A noticeable difference with Courts of last resort in typical common law jurisdictions is that the Supreme Court has no discretionary power to select its own caseload: all civil—and criminal—judgements of trial Courts are subject to appeal.[63] It thus also performs the function of a common law intermediate appellate Court. Moreover, the Supreme Court may in certain instances review facts and even rehear evidence not unlike that of a continental Court of appeals that exercises 'trial' jurisdiction (*juridiction du fond*).[64] This

---

[59]Administration of Justice Law 1964, art 3(4). On President Zekia see Hadjihambis (2010), pp. 94–97. Zekia became also the first Cypriot judge at the European Court of Human Rights, from 1961 until his death in 1984.

[60]A corollary speculation concerns the possible orientation of Cyprus public law within the Continental legal tradition: the departure of Forsthoff led to the monopolisation of Continental public-law influences by the Greek administrative law tradition, which at the time was strongly oriented towards the French.

[61]Papasavvas (1998), pp. 127–144; Kombos (2015).

[62]In the sense suggested by Kennedy (1997).

[63]Courts of Justice Law 1960, arts 25(1) and 25(2).

[64]Ibid, art 25(1)(3): 'Notwithstanding anything contained in the Criminal Procedure Law or in any other Law or in any Rules of Court and in addition to any powers conference thereby, the Supreme Court, on hearing and determining any appeal either in a civil or a criminal case, shall not be bound by any determinations on questions of fact made by the trial court and shall have power to review the whole evidence, draw its own inferences, hear or receive further evidence and, where the circumstances of the case so require, re-hear any witnesses already heard by the trial court, and may give any judgment or make any order which the circumstances of the case may justify...'.

power—introduced immediately upon Independence—has been held however to apply only in extreme cases.[65]

What all this means, however, is that only a fraction of the appeals pose real legal questions. The Justices are thus left with relatively little time on their hands for serious research. Some Justices, in some cases, effectively tend to simply choose between arguments presented by counsel.

10.2.1.3 Administrative Jurisdiction

An important reason for the day-to-day influence of the Supreme Court and perhaps the single most important factor to its overloaded docket lies in the Court's trial-level jurisdiction over administrative law cases under Article 146 of the Constitution. In the spirit of mid-twentieth century Continental administrative law, such jurisdiction is only limited to annulment of administrative acts, as opposed to administrative litigation *au fond*. Be that as it may, there are a lot of administrative cases—infinitely more than what the drafters of the Constitution had in mind when they assigned them to the then Constitutional Court. As a result, until the creation of a first-instance Administrative Court a year ago, the Supreme Court justices had been spending more than half of their time—and almost all the time of the Court's legal officers (law clerks)—judging administrative cases individually (an arrangement colloquially referred to as 'single bench'). Whereas civil and criminal appeals are examined by three-member panels, appeals against Supreme Court trial judgements are considered by five Justices other than the one at first instance: such a panel is characteristically, if rather confusingly, called a 'plenary bench' in the colloquial legal language of Cyprus. The entire Supreme Court may be called upon in cases of great interest.[66]

Requiring a senior judge, who has spent decades to reach appellate Olympus, to actually adjudicate *en masse* small trial-level cases would be hard on anyone from

---

[65]Courts of Justice Law 1960, arts 25(1) and 25(2), art 14. For the state of law during the British colonial period, see *Charalambous v. Demetriou* (1961) C.L.R. 14 (the last case decided under the final colonial-era Courts of Justice Law (Cap. 8). In the words of the Court's reporter, the opinions ('judgments') of the three Cypriot High Court Judges contain 'a restatement of the powers of Appellate Courts in Cyprus under the old law in disturbing findings of fact of trial courts' For the restrictive stance of appellate case law, see, e.g., *K.K. v. Attorney General* (2008) 2 C.L.R. 294, *Kleitou v. Republic* (2011) 2 C.L.R. 113, *Attorney General v. Arabides* (2013) 2 C.L.R. 113. For an actual case of appellate intervention into factual findings see the very recent civil case *Xenophontos v. Zoo Bar Restaurant Ltd*, judgment of 15 December 2016.

[66]See *Christodoulou v Public Service Commission* (2009) 3 C.L.R. 164, [2010] 3 *Lysias* 116. The full Court sat on first instance in this petition by a Supreme Court Registrar to annul the appointment of another colleague to the position of Chief Registrar of the Supreme Court; the *Christodoulou* Court accepted the petition, setting formal standards and limits on how much the interview of candidates for appointment or promotion to public service may be taken into account over formal qualifications. The President of the Court had recused himself.

any legal system. But here all Justices have spent the better part of two decades or more sitting on anything but administrative cases. A lot of them have never studied administrative law prior to ascending to the Supreme Court bench. These circumstances have led to a stronger role for the Court's law clerks or legal officers, who hold permanent positions as assistants to individual judges and tend to have studied in Greek law schools.

In 2015, after decades of discussions between bar, judiciary and legislators (many of them members of the legal profession), reform finally occurred. Even though most jurists with an opinion on the matter suggested that a return to the original constitutional arrangement of separating the Constitutional Court from the High Court may have been the most proportional solution, government and legislature adopted the position promoted by the judicial establishment: the unified Supreme Court remained intact and a subordinate Administrative Court, based in the capital, was created to take over first-instance jurisdiction over administrative cases, including tax litigation.[67] Since early 2016, most administrative cases are being judged by a single Administrative Judge and a three-member Supreme Court panel on appeal.

10.2.2 The Legal Profession

10.2.2.1 Definition

Cyprus follows the common-law notion of a single, Bar-centred *legal profession* (νομικό επάγγελμα), as opposed to the plural number (νομικά επαγγέλματα) often used in Continental countries such as Greece.[68] The common law tradition is keener on a unitary conception of the legal profession—presumably a result of the provenance of common law judges from the Bar, the historical weakness of career government lawyers separate from the Bar and the lack of distinct legal professionals with specialised tasks or training, such as notaries.

The English technical term used to define a member of the Cyprus Bar is *Advocate*: the term, also used in Scotland and former British colonies from South Africa to South Asia, has been carried over from the colonial period. In the Greek language—i.e. the official language of the Republic—the term used is the same as in Greece: δικηγόρος. Ironically, even though the term *advocate* has found its way to official translations of the Greek term by at least some Bar Associations in Greece, there is a tendency in common language in Cyprus to translate in reverse the term δικηγόρος as 'lawyer'.

---

[67]See the Establishment and Operation of Administrative Court Law 2015 (L. 131(I)/2015); L. 130 (I)/2015, passed with a super-majority, had immediately prior amended the Constitution, adding a paragraph- 146(1A) - to Article 146 which enabled the creation of an Administrative Court.
[68]European Justice (2016).

The Law regulating *advocates* is the second chapter in the colonial collection of the *Laws of Cyprus*.[69] The Law's description of what constitutes 'practicing as an advocate' (ασκείν την δικηγορία) includes both litigation-related tasks and the basic forms of consultative lawyering.[70] The traditional English split between barristers and solicitors appears, therefore, alien to Cyprus and appears to have never formally existed in the island's legal profession. In practice, however, Cypriot advocates often present themselves as 'lawyers and legal consultants' in business cards and storefront displays of law firms.

The structure of the Cyprus legal profession is in that regard indicative of the forces at play. In many of the former British colonies, advocates, just like barristers, coexist with solicitors. In fact, the original colonial-era definition of 'practicing as an advocate' included only Court-related tasks.[71] The fact that trained barristers tended to dominate the colonial bar further strengthened this connection between being a *lawyer* and appearing before the Court.

After Independence, Article 2(1) was gradually amended and the definition expanded. Most consulting services presently enumerated having been added in the early 1980s. This still leaves outside, however, legal officers (νομικοί λειτουργοί) in the civil service, as well as lawyers employed in-house by a business on a fixed-salary basis (έμμισθοι δικηγόροι). This creates problems: members of these groups cannot enjoy effective professional representation; their experience does not count towards the statutory requirements for judicial office. They are also excluded from participation in the Advocates' Pension Fund—in fact, the desire to keep them out of the Fund may have been an important motivation for keeping them out of the Bar itself. Given the strength of big auditing/consulting firms in Cyprus, including the Big Four, many in the profession are also weary of a possible encroachment on legal subject matter by accounting offices.

10.2.2.2 Organisation

Cypriot advocates are organised into the Cyprus Bar Association, which constitutes the countrywide licensing body.[72] They are also enrolled into a local Bar Association (one for each of the original District Courts), which takes charge of day-to-day affairs.[73] The largest association is Nicosia, followed by Limassol.

---

[69] See the Advocates Law Cap. 2 (L. 58/55; 'A Law to consolidate and amend the law relating to advocates and to make provision for the establishment of an Advocates' Pension Fund', as amended by L. 24/56). The Law has been amended over 30 times in the 50 years since independence; [hereinafter Advocates Law].

[70] Advocates Law Cap. 2, art 2(1) as amended.

[71] The original Article 2 provision of L. 58/55, made reference only to Court-related tasks. See Trusted (1959).

[72] Advocates Law Cap. 2, arts 21–25.

[73] Ibid., arts 19–20.

The President and board members of the district Bar Associations are elected directly by their members. The Board of the Cyprus Bar Association is composed by representatives of the district Bar Associations, some of which participate ex officio but some are elected directly by the district Bar assembly. The President of the Cyprus Bar Association is elected directly by those members taking part in a countrywide electoral assembly; given the fact that this assembly is held in the capital Nicosia, home to the largest district bar, Nicosia tends to dominate.

Requirements for admission to the Bar include a law degree, pupilage for at least a year with an advocate, and success in exams organised by the Law Council—which consists of the leadership of the Cyprus Bar Association, the Attorney General, and experienced advocates selected by them.[74] Once admitted to the Bar, Cypriot advocates are allowed to present themselves before any Court throughout the Republic.

Advocates play a strong role in politics—partly for the usual reasons and partly because of the strong legalist discourse concerning the Cyprus problem. Parliamentarians are free to appear before Court—leading to an ever-stronger role for the legal profession as a pressure group.

10.2.2.3 The Evolving Demographics of the Cyprus Bar

The composition—and background—of the Cyprus Bar is representative of the evolution of the legal system.

First, with regard to composition and numbers, the Cypriot legal profession was kept in relatively small numbers well after independence—when access to the law schools of especially Greece allowed children from the lower middle class to pursue legal careers. The numbers begun to increase noticeably in the 1980s and 1990s, to what could still be regarded as manageable levels. The expansion became dramatic in the years immediately prior and following accession to the EU, with a veritable wave of UK-educated new entrants being added to the existing streams. The launch of a law program at the national University of Cyprus and especially the creation of several private law schools has led to an explosion in numbers, at least by Cyprus standards: we are now at 3000 members of the Cyprus Bar Association, but those taking the bar exam each year are now in the several hundreds or more (from an average of four dozen twenty years ago). The gender balance has also been transformed: the first woman advocate (coming, as may be expected, from a dynasty of well-connected lawyers) was only admitted to the Bar in 1951; today women represent just over half of Bar membership but a very strong majority of new entrants.

Second, and perhaps more importantly, with regard to background. Prior to independence, especially following the 1931 revolt, members of the Bar—including government lawyers and the judiciary—were trained in England and Wales

---

[74]Advocates Law Cap. 2, arts 21–25, art 3.

(often without completing or even beginning university education in law). In fact, according to biographical data, the majority of native lawyers admitted to the profession prior to 1931 had held university degrees in law—primarily from the University of Athens Faculty of Law. Following the revolt, successful training in the United Kingdom as a barrister (or, less frequently, a Scottish or colonial advocate) became a prerequisite—the sole prerequisite.

After independence in 1960, the majority of new entrants into the profession had obtained university degrees from Greek law schools; the United Kingdom remained the destination of choice for a minority, which included, however, most of the sons (and, gradually, the daughters) of the colonial-era Greek Cypriot barristers. Continental concepts and terms were introduced into Cyprus law, but less than might be expected in terms of the Bar's demographics. Moreover, it took more than three decades for English to be replaced by the Republic's official languages in Courts (and colonial statutes to be translated). Both these phenomena could be explained in terms of a contest between the various generations and social groups constituting the (Greek) Cypriot Bar.

The post-colonial character of the legal system as well as the lack—until very recently—of a legal academia have also meant that the Bar has remained deferential to the judiciary much more than might be the case in other European countries.[75] There still exist relatively few publications on Cypriot law; even though their numbers have increased in the past few years, the bulk still tends to be limited to the uncritical presentation of basic local case law. It is difficult for a practising advocate to be critical of Court reasoning on record. Public deference to the judiciary has been traditionally very strong, even if the Bar Association has on occasion taken a different stance from judicial leadership in certain proposals involving law reform. In a small world like the core Cyprus legal milieu, the social interaction (and kinship ties) between members of the judiciary and members of the bar is more frequent than especially in bigger, Continental countries—a fact that helps both to enforce deference and forge consensus.

10.3   Law in a Biggish Small State

10.3.1   The Persistence of English Law: Contract and Commercial Law

Contract and commercial law—in the English sense—is a good case in point: Cyprus law is effectively English law in these subjects. Even after independence, new legislation such as the 1994 Sale of Goods Act simply replaced one English

---

[75]Even within the small world of Cyprus legal academia, it is not common for teaching staff (especially those employed in the private law schools, who tend to have a strong professional or personal interest in legal practice), to be critical of Court judgements, at least in public.

statutory transplant (the Sale of Goods Act 1893) with another (the Sale of Goods Act 1979).[76] Local case law does exist—and deviations from the common law are not completely unheard of—but the number of landmark cases is relatively small. Moreover, the reasoning in most Cypriot contract cases is certainly not a match for appellate English opinions written often by specialised judges with time—and economies of scale—in their hands. A reputed English reference work like Chitty is thus often a more readily available resource for stating the law than a search into local cases.

The hold of English contract and commercial law is such that even legislated deviations from the English paradigm—such as the Vienna Convention on the Sale of Goods or statutes implementing EU directives—are only absorbed very gradually in legal practice and Court decisions. For example, more than 10 years after the Vienna Convention entered into force in Cyprus, no case law exists on the Convention. Indeed, even the country's legal literature makes little reference to the CISG as being in force in Cyprus.[77] European derivative law is just beginning to make inroads into case law and contractual practice: the Unfair Contract Terms Directive was already implemented prior to EU accession, with the Unfair Terms in Consumer Contracts Law 1996—but only very recently have we begun to notice Cyprus cases enforcing the prohibition against unfair terms and exploring issues settled in the case law of other Member States. Effective consumer protection is still in its early stages.

Be that as it may, the biggest challenge to the hold of English contract and commercial law in Cyprus comes not from native doctrinal thinking but from other authoritative external sources of law reform—the European Union or international instruments.

10.3.2   Mutation in the English Core? Civil Procedure

On the face of it, procedural law in Cyprus falls firmly within the common law tradition; in fact, as was already noted in several occasions above, even in fields where substantive law is oriented towards the Continental legal tradition, the rules of procedure are a principal vehicle for the introduction of common-law notions—and for ensuring the persistence of a common-law mentality.

---

[76]Sale of Goods Law 1994 transplanted the English 1979 Sale of Goods Act; it repealed the Sale of Goods Law (Cap. 267), enacted as L. 25/1953, which had transplanted the 1893 English Sale of Goods Act; the 1953 Law had in its turn repealed the chapter on the sale of goods in the Contract Law (Cap. 149), which was modeled after the 1872 Indian Contract Act.

[77]For example, no reference is made to the CISG being in force in Cyprus in the main English-language reference work on Cyprus law, authored by a leading law firm: Malliotis (2010), pp. 749–802.

A closer examination confirms the primacy of the English legal tradition in procedural matters: whereas in Continental counties the principal source of procedural law lies in comprehensive Codes of Civil Procedure enacted by legislation, in Cyprus legislative texts come second in importance to rules of procedure enacted by the Supreme Court. The Courts of Justice Law 1960, which should be regarded as the starting point for any litigation, addresses matters of venue and subject matter jurisdiction and sets the principal rules on the status of judges and the function of the Republic's Courts. The so-called Civil Procedure Law (Cap. 6), which goes back to the first period of British colonial rule, addresses primarily matters of enforcement. The Evidence Law (Cap. 9) contains an interpretation clause designating as default legal framework in evidence matters the law of England as of 1914.[78] Cyprus has adopted the UNCITRAL Model Arbitration Law,[79] but domestic arbitration is still governed by the Arbitration Law (Cap. 4), which has shared roots with the English Arbitration Act 1950. With the exception of part of the Evidence Law,[80] there has been very little reform of procedural law—whether legislation or Court Rules—since independence. Indeed, it took over 30 years since independence for the official translation of colonial-era statutes such as the Civil Procedure Law and the Evidence Law into Greek, whereas there has so far been no official translation of Court rules, including the flagship Civil Procedure Rules; only new Rules or amendments to the Rules are enacted in Greek.

These same factors, however, have also ensured a differentiation between English and Cypriot law. Cyprus procedural law is English law—but English law from a time vault. Cyprus law did not follow the sweeping reforms in English civil litigation or arbitration that occurred especially since the 1990s. As a result, Cyprus judges and legal writers often refer to English reference works addressing UK law prior to the 1999 reform of civil procedure or prior to the 1979 Arbitration Act respectively. The growing distance from overseas authorities and the lack of substantial procedural reform has led Cyprus litigation practice to develop inwards: the day-to-day needs of Courts and practitioners have to be catered to, even by the haphazard creation of local precedent. This quasi-spontaneous process of precedent creation is not without its problems: even the question of whether misstating the correct legal basis of an interim application was a ground for dismissal has led to competing lines of precedent between Supreme Court panels.[81] It is also not universal: modern developments in English commercial litigation are still a guiding light, especially when it can be argued that they are based on—or at least are compatible with—Cyprus procedural rules.[82]

---

[78]Evidence Law (Cap. 9), art 3. For an explanation see Hatzimihail (2015), pp. 73–74.

[79]International Commercial Arbitration Law 1987 (L. 101/87).

[80]Evidence Law (Cap. 9), arts 23–36 as introduced by L. 32(I)/2004.

[81]Hatzimihail (2015), pp. 96–97.

[82]Ibid., pp. 94–96 examining the development of Cyprus Supreme Court's attitudes toward worldwide freezing orders.

10.3.3   Civilian Enclaves and Their Mutation (I): Family Law

When the British took over Cyprus from the Ottomans, acknowledged ethnic-
religious groups had, under the *millet* system, their own laws and tribunals to
determine matters of personal status—essentially issues classified within family
and succession law today. The so-called *personal laws* system still survives across
the eastern coast, notably in Lebanon, Israel and Syria. In the case of Cyprus, where
the vast majority of the population was Greek Orthodox, this bestowed institutional—as well as political—prominence on the Church of Cyprus, which applied
Roman-Byzantine law to Greek Cypriots. Apart from the Moslem (Turks), other
ethnic or religious groups included the Armenians, Maronite and Latin (Roman
Catholic).

The British colonial administration gradually but drastically encroached upon
this ecclesiastical jurisdiction: questions of marriage and divorce were left with the
religious communities, but anything of pecuniary interest—i.e. including testate
succession, probate proceedings, marital property and custody—was taken over by
the colonial judicial system and British colonial legislation.[83] Given the colonial
quasi-monopoly in adjudication, Anglo-colonial law begun operating in parallel
even to the core of personal law. For example, actions of breach of promise to marry
started to supplant, before the District Courts, local customs and the Roman-
Byzantine law on engagements. In the Turkish Cypriot community, Islamic family
law had disappeared by the end of the colonial era, substituted by the transplantation of secular law from the Republic of Turkey. Independence in 1960 stabilised
things. Article 111 of the Constitution guaranteed ecclesiastical jurisdiction over
matters of engagement, marriage and separation and 'family relations' but the
colonial status quo remained until Article 111's amendment and the creation of
secular Family Courts in the Republic.[84]

In 1990, as part of a comprehensive reform of family law, the Republic
established secular Family Courts.[85] The Family Courts were originally intended
to replace ecclesiastical tribunals, especially with regard to Greek Orthodox
Cypriots.[86] Over the past two decades, the jurisdiction of the Family Courts has
been expanded, by statute and via the case law of the Supreme Court, both *ratione
materiae* on every aspect of family law, and *ratione personae*. At the moment, the
main exceptions concern the validity and dissolution of marriage under the rules of

---

[83]See the Courts of Justice Law 1953 (L. 40/53, Cap. 8), art 34. For English-era family legislation
see Caps. 274–280 in the Statute Laws of Cyprus, vol. 5.

[84]For an overview of things prior to the family law reform see Serghides (1988).

[85]Family Courts Law 1990 (L. 23/90), as amended.

[86]In fact, Family Courts were originally supposed to be presided over by a cleric, flanked by two
lawyers, but the Church of Cyprus refused to participate. What did remain from this effort to
accommodate the Church over its loss of constitutional privilege is the requirement of three-
member panels sitting on divorce petitions, the overwhelming majority of which pose no real
dispute (as opposed to the nastier separate trials on children and especially marital property).

the three religious groups (which fall under the jurisdiction of the respective Family Tribunal),[87] and some cases involving Turkish Cypriots.[88] A three-member panel of Supreme Court justices, rotating in two-year terms, sits on appeal as the 'Second-Instance Family Court'.[89]

Judicial reform provided the impetus for a secularisation of Cypriot family law. Remarkably, this secularisation brought in a sense Greek Cypriots closer to their roots than the 'English' regime from colonial era. In a series of statutes on marriage, divorce, marital property, paternity and children care, the Republic of Cyprus effectively transplanted the bulk of Book IV of the Greek Civil Code, which had been recently reformed in a secular direction taking account of gender equality and establishing no-fault divorce.[90] The only subject that was not Hellenised (or, in a way, re-Hellenised) was adoption: partly because of the drafter's background, partly because this is a subject involving social services and hence triggering bureaucratic inertia, perhaps because adoption was initially expected to fall within the jurisdiction of the 'common law' District Courts, but most likely because that was a subject where reform was still debated in Greece so there was no modern model to emulate.

As a result, we have in family law a legal field in which substantive law is, on the face of it, purely continental—a case of the effective transplantation of modern Greek family law. This is commonly acknowledged and, in a way, perpetuated. All but one of the ten family Court judges as of 2015 were educated in continental law schools (that is, in Greece). This also holds true of most attorneys appearing regularly before the Family Courts. Family law textbooks from Greece present strong authority for the Family Courts and even on appeal. The Continental flavour of the field—and the availability of easily accessible material in Greek—may account for the comparative proliferation in local writings on family law, mostly by senior family judges.

Be that as it may, Cyprus family law is a true hybrid. Part of its hybridity is a matter of procedure. Processes before Family Court retain much of the common law air of Cyprus litigation; common law institutions such as cross-examination of witnesses are used by Family Judges (alongside inquisitorial techniques not available to their District Court brethren).[91] Whereas in Greece a single divorce trial

---

[87]Separate Family Tribunals of Religious Groups were also set up to deal with the divorces of members of the three religious groups (Armenian, Maronite, Latin) recognised by the Constitution: these tribunals are composed of one 'President' judge, appointed by the Supreme Court 'from among members of the judicial service', one District Judge, and one representative of the respective group. See the Family Courts (Religious Groups) Law 1994 (L. 87(I)/94), especially art 3.

[88]See Serghides (2010), pp. 24–29.

[89]Family Courts Law 1990, art 21(1).

[90]See Nicolaou (1998), pp. 121–34.

[91]The recent attempt by the Supreme Court justices—who are tasked with drawing and revising rules of procedure for all Courts—to reduce the oral part in certain types of family cases provoked an uproar by the family bar.

takes care of all questions arising out of marriage dissolution, and a non-contentious process of 'consensual divorce'—where the judge signs off on the spouses' settlement provided they did reach agreement on all critical matters—is also available, in Cyprus there is no consensual divorce and a marital dispute may involve four separate Court cases—one each for divorce, marital property, child support and family home (which has the adverse result of making the overwhelming majority of divorce trials effectively non-contentious, but also of prolonging marital property litigation both before and after divorce is sought). If this does not seem to impact the common law/Continental divide, appellate control by Supreme Court justices educated mostly in Britain and with little academic or practical experience in family law probably does.

In a sense, the influence of Greek family law is relatively shallow: leading cases will cite Greek textbooks, but references to Greek family case law are less common. As local practitioner-oriented handbooks develop, the textbook influences from the mainland are becoming more instrumental and piecemeal. On the other hand, monographs and specialised scholarly contributions are very seldom cited or consulted—which is unfortunate given that their doctrinal and policy discussions may be more pertinent than textbooks and in a language understandable by Cyprus family lawyers. The fact that this is a niche field, with a finite number of judges and practitioners, does not help this envisaged enrichment of family law.

10.3.4   Civilian Enclaves and Their Mutation (II): Administrative Law

The British left Cyprus in 1960 with the foundations of an effective civil service, but almost nothing in terms of administrative law. In fact, judicial review of administrative acts was at the time a novel subject in Britain, with the first notable works of reference only just appearing and certainly not having had an impact on English-trained Cypriot barristers. Reserved for civil and criminal matters, the common law was not supposed to fill this void in the new Republic. This was the job of the Supreme Constitutional Court, presided by a Continental public law academic and tasked with administrative jurisdiction. Had the Court remained intact, it would have probably developed in the spirit of Continental Supreme Courts on public law, such as the Greek Council of State. Ironically, such a Supreme Constitutional Court with German and Turkish input would have had to develop in a more autonomous fashion from Greek administrative law as well—probably creating an arena for competition between the French and German public law traditions and even feeding back into the public-law discourses of Greece or even other countries.

But this was not to be. The Supreme Constitutional Court was merged, effectively absorbed, into the 'common law' High Court. The new Supreme Court consisted exclusively of Cypriot British-trained barristers in origin. Most were also Greek in ethnicity, even though it took another 30 years for the first judge educated in a law school in Greece to make it to the Supreme Court bench and sit on administrative cases. Greek law did become the dominant force in the development of Cypriot administrative law. The case law of the Greek Council of State was even called 'nourishing mother' of the Supreme Court one.[92] The main treatises from Greece are still cited—indeed the General Principles of Administrative Law Law 1999, which codified principles and Cypriot case law relied on treatises by Athens professors (and a draft Code of Administrative Procedure prepared by yet another Athens professor, a long time prior). Administrative litigation is a privileged field for lawyers educated in Greece. Indeed, of the five appointees to the new Administrative Court all were educated in Continental law schools (all but one in Greece). So are almost all legal officers (permanent law clerks) that research cases and often prepare drafts for Supreme Court justices sitting on administrative cases. Administrative law is the Continental enclave par excellence, the basic reason to put a dot on Cyprus on the comparative-law map.

On the other hand, administrative law has been until now entrusted to judges who, in their vast majority, had no university education or vocational training on the subject, had practised little or no administrative law prior to their elevation to the bench and had had no contact with administrative cases in the decades they spent on the District Court (and this will remain so for administrative appeals, perhaps for decades). Lack of background, workload pressure and often a slight condescension are not always helpful in constructing a system of administrative law.

10.4 Legal Elites and 'Gatekeepers'

The degree of mutation in family and administrative law, including legislation inspired by Greek law, is stronger than with English law transplants. Which begs the interesting question why. Legal education is certainly not an important factor, at least up to the present, although in the long run it will alter things. National identity is also not a factor. As to the relationship between Cyprus law and society, it has been one of co-existence, 'live and let live' rather than a dynamic one of mutual influence. Nor does it help much to think in terms of a market for, or competition between, transplants and influences. Perhaps there is something to the idea of 'dominant' legal fields, such as contract or tort, that set the tone of the legal system. Moreover, it probably takes more effort at adaptation in order to fit the Continental transplants into the existing legal system than the English ones.

Another important factor has to do with the power politics within the legal profession—especially the role and composition of the legal elites.[93] It could be

---

[92]*Zittis v Republic* (1998) 3 CLR 394.

[93]Symeonides (2003), pp. 449–450. Symeonides has been harsher, speaking of 'Anglicization' following independence and pointing the finger at members of the (Greek Cypriot) colonial elite who tricked an 'inexperienced' legislature into tying Cyprus law 'surreptitiously and permanently' to the English common law.

argued that the most influential groups within the legal profession—appellate judges and the *notables* such as families that trace their origins in the profession to the colonial-era advocates or the first years after independence—are the ones most closely identifying with the common law. By consequence, these groups have had a vested interest—and the position—to act as gatekeepers and to remain reluctant to either learn new things or to open up the system to new entrants who have studied in Greece. This has certainly been the case on quite a few occasions—including delaying for over 30 years the translation of basic statutes into Greek, as mandated by the Constitution and the continued use of English in Courts for almost that long, despite it having been abolished as an official language by the constitution in favour of Greek and Turkish.

Gatekeepers exist in all systems—in a small state, however, it is more difficult to disperse what was recently described by one of the most notable among *notables* as 'legal power' (νομική εξουσία), as he called for its separation from 'political power'.[94] I would identify three groups of gatekeepers, represented by the upper echelons of the judiciary, the Law Office of the Republic and the Bar establishment.

10.4.1 Judiciary

Partly by the power of law and partly by the force of necessity, the judiciary has been endowed with powers not unlike those of judges and justices in a common law jurisdiction. Cyprus judges enjoy the respect of Cyprus society; however, they are often defensive of their status and do not tolerate challenges from either advocates or the public. The notion of contempt of Court was used expansively,[95] until the European Court of Human Rights called Cyprus to task.[96] However, the judiciary

---

[94]Intervention by phone by Polyvios Polyviou, Advocate in a live television news show at the Cyprus Broadcasting Corporation, on 14 April 2015, in which there was talk of mutual accusations between the Attorney-General and Assistant Attorney-General (which resulted in the prosecution of the latter by the former and legal action to strip him of his office). The President of the Republic had considered appointing a special counsel, independent to the Law Office, to examine all claims but was swiftly forced to reconsider following reaction by the bulk of the legal establishment. All allegations by the Assistant Attorney General were then dismissed by prosecutors. The Attorney-General was subsequently able to secure the impeachment of the Assistant Attorney General by the Supreme Court. The criminal case is, as of January 2017, awaiting judgement by the Assizes Court.

[95]Courts of Justice Law 1960, art 44. The original article effectively reprised art 49 of the Colonial Courts of Justice statute 1953 (L. 40/53, Cap. 8). It was amended significantly.

[96]See *Kyprianou v Cyprus* (G.C.), 2005-XIII Eur. Ct. H.R. (Appl. no. 73797/01). The Supreme Court of Cyprus, *In Re Kyprianou* (2001) 2 C.L.R. 236, had upheld the conviction of an advocate by the Assizes Court of Limassol for complaining that the judges on the bench were exchanging 'billets doux' during his speech.

has exercised notable self-restraint in matters of political sensitivity.[97] Even in less political subjects, we will search in vain for systematic efforts by the appellate bench to reshape the law. In fact, the recent tendency in many landmark cases appears to be to avoid expansive reasoning.

Several interconnected factors have led to political autonomy and a particularly strong role for the Cyprus judiciary, especially its upper echelons, including: the existence of only one trial instance; the lack of an intermediate appellate Court and especially the merger, a few years after independence, of the High Court and the Supreme Constitutional Court into one single Supreme Court, which also sits as the Supreme Judicial Council that appoints and promotes judges; the consensus politics of a coalition-based presidential system; the effective inability at constitutional reform; last but most certainly not least, the creation in practice of a self-perpetuating, hierarchical judiciary with seniority being the decisive factor of promotion to the upper echelons has led to political autonomy and an especially strong role for the judiciary.

As might befit a small state with a common-law system and limited decentralisation of the administration of justice, Cyprus judiciary is not big in numbers.[98] It is nonetheless a hierarchical group, in which there is a discernible core and a periphery, as well. The judicial elite is constituted of the 13 Supreme Court justices and the senior judges at the district Courts, from whose ranks virtually all Supreme Court justices have come since independence.

10.4.1.1   Controlling the Entrance to the Judiciary

To be a judge in Cyprus means embarking upon a judicial career, which despite the common law attitude shares many affinities with continental models of a hierarchical, career-based, even corporatist judiciary.

According to the statute books of Cyprus, not unlike other common law jurisdictions, judicial appointments come on the basis of a successful career in the legal profession, with direct appointment to the higher ranks of first-instance judges, or even the Supreme Court, being possible.

There are three ranks of District Judges: District Judge, Senior District Judge, and President of the District Court.[99] What is required for an entry-level

---

[97]See *Kettiros v Koutsou* (2007) 1 CLR 828, [2008] 1 Lysias 71 with editors' note: even though the law on parliamentary elections effectively penalises coalitions of parties as opposed to single party lists, the Court unanimously held that it is a matter for the electoral list itself to define its status.

[98]At present there are 79 judges in total at the District Courts, of which 48 at the entry-level position of District Judge, 17 at the intermediate level of Senior District Judge and 14 at the senior position of President. The Courts of Justice Law, art 6(3) sets maximum numbers for each rank (these were increased by L. 41(I)/2014, from a previous maximum of 39, 16 and 13 respectively). There are also three Presidents of Family Court (each for one of the Family Courts) and seven Family Judges.

[99]Ibid., art 4.

appointment to the District Court is 'high moral standards' and a minimum of 6 years in legal practice; 10 years in legal practice are required for direct appointment to the middle and senior ranks of first instance.[100] The legal practice requirement can be fulfilled by 'service in any judicial position'.[101] This includes the staff of the Law Office of the Republic (who even possess an advantage over private practitioners, since they can be immediately appointed to any District Court, whereas the latter may not be appointed to the District Courts related to the territory of their local bar associations), as well as advocates employed by the Supreme Court as reporters or editors of Court decisions.[102] On the contrary, it does not include legal officers, including the Supreme Court's own clerks. The practice requirement can be reduced to 5 years for entry-level appointees, on the advice of two thirds of the Supreme Court Justices.[103]

New judges must have therefore been members of the Bar in good standing for the appropriate period of time prior to their appointment, but, unlike the English Courts of justice, they do not really have to have distinguished themselves before the Courts (academia is and will long remain unthinkable as a testing ground). The selection process—operated by the Supreme Court Justices, in their capacity as the Supreme Judicial Council—principally involves an interview. One might be tempted to say that neither the safety valves of continental systems—exams, judicial training—nor those of common law systems—reputation among the Bar and the legal profession—are in place. However, the Cypriot legal profession is a small world and reputations are easily confirmable: moreover, the senior judge from the applicant's district Court is expected to provide a reference for the applicant. In fact, getting qualified candidates to apply has often been a bigger problem than selecting the best suited among those who do apply. On the other hand, there has been increased interest in new positions—probably as a result of the explosion in lawyers' numbers combined with the economic crisis of the past few years, which have been making private practice less rewarding financially than in the past.

Hierarchies within the judiciary are displayed even at the entry level. Appointments to the new Administrative Court must fulfil the requirements for senior district judges, i.e. 10 years in legal practice, although the requirement may be reduced to 8 years for entry-level Administrative judges.[104] 'Broad knowledge in administrative law matters or proven experience in handling court cases falling within the jurisdiction of the Administrative Court' is also required. In the fall of 2015, however, when the first President and Administrative judges were appointed, the legal practice requirement was invoked to exclude the legal officers of the Supreme Court from the final round of the process, despite their acknowledged

---

[100]Courts of Justice Law 1960, art 6(1).

[101]Ibid.

[102]Ibid., art 6(1). This provision was added by L. 35/82 and was successfully invoked only once—soon after this amendment was enacted.

[103]Ibid., art 6(2).

[104]Ibid., art 4.

experience in handling administrative litigation for many of the same justices who constituted the Supreme Judicial Council. Applicant judges were also excluded and the decision was made to appoint private practitioners engaged in administrative litigation, as well as one staff member of the Law Office. On the contrary, bar membership (or even bar admission) is not formally required for appointment as a Family judge. As a result, even though most family judges came from the legal profession, the Supreme Judicial Council has on occasion appointed legal officers of the Court—one was even made President of the Family Court soon after its creation.

10.4.1.2 Hierarchies Within the Judiciary

Once appointed, trial judges are scrutinised from the higher judicial echelon, not unlike in continental systems. Trial judges are dependent for their promotions and transfers between districts on the 13 Justices of the Supreme Court, who act as the Supreme Judicial Council. Disciplinary proceedings before the Council are not unknown. Internally, each District Court is presided over by the senior President (known in the colloquial legal language as the 'administrative President'). Family and Administrative Courts only have one President: given the handful of judges in each, administration is a more casual affair but still under the full control of the Supreme Judicial Council. Even the organisation of judges to a Union[105] tends to strengthen this hierarchical structure (as well as the elite mentality in the core): the Union tends to elect as its president a senior President of the District Court, likely to ascend to the Supreme Court bench.

In contrast to entry-level appointments, appointments to intermediate and senior ranks has been exclusively a matter of internal promotion, with a strong emphasis on seniority. 'Sorting out' takes place in the first two ranks, i.e. among District Judges and to a lesser extent among Senior District Judges. The Supreme Judicial Council has some leeway in promoting Senior District Judges to Presidents of the District Court, but it tends to respect seniority at this level.

Supreme Court appointments are almost strictly a matter of seniority—and judicial promotion. The Constitution, in fact, provides that appointment to the appellate bench is made by the President 'from amongst lawyers of high professional and moral standard'.[106] However, all but one appointment since independence were made from the ranks of Presidents of the District Court. The only appointment made from outside the ranks of senior judges was in 1997, when, heeding calls from the Bar for an advocate to sit on the appellate bench, the then

---

[105]Creation of the Union was enabled by art 4 of L. 136/91, which also added art 10A to the Courts of Justice Law 2960.

[106]The Constitution of the Republic of Cyprus, art 153(5).

President of the Republic, on the advice of his Attorney General, named a senior prosecutor to the Court in one of the two openings, prompting a violent reaction by the district judges.[107] This became the proverbial exception that settles the rule: when a vacancy arises, the President will name the most senior President of District Court to the Supreme Court. Likewise, the Supreme Court's most senior justice is to become President once the previous one retires.

There is a lot to be said about this system of rigorous seniority. It certainly acts as a disincentive for experienced lawyers to enter the judiciary at a more mature age. On the plus side, it fosters a very strong esprit de corps within the judiciary: whereas this may on occasion lead to a bunker mentality, the potential for friction and competition among the judges is also drastically reduced. Most importantly, the capacity for political intervention and patronage is minimised.

The status of Family judges within the family of Cyprus judges offers another aspect of the hierarchies and structures of the judicial elite. Chances of promotion are minimal: there can only be one President in each Family Court so one of the Presidents has to retire or step down for a Family judge to be promoted. Further promotion is at best even more difficult: Supreme Court justices have long been opposed to the idea of a President of a Family Court being eligible for appointment to the Supreme Court. By now, a precedent has been created: when several openings became available, a few years ago, some of the Presidents of District Court elevated to the appellate bench had both fewer years in judicial service and significantly fewer years as Presidents than the most senior President of Family Court, who also boasted of academic experience and publications in matters such as administrative and human rights law. In fact, Presidents of Family Court are equated in salary with the Senior District Judges, rather than the Presidents of District Court, and do not enjoy certain additional financial benefits of their general-jurisdiction brethren.

Another source of power for the judicial elite comes from appointment to international positions, namely the European Court of Human Rights and the Court of Justice of the European Union in Luxembourg (a President of the Supreme Court was even nominated to the International Criminal Court). Supreme Court Justices and, on occasion, senior district judges have had a virtual hold on these positions, challenged only by the upper echelons of the Law Office the Republic.[108]

---

[107]Hellenic Resources Network (2016). The appointment was condemned by the Union of Judges in their general meeting, with 34 (out of 44) Judges present tending their resignation.

[108]The election, in early 2016, by the Parliamentary Assembly of the Council of Europe, of Dr. mult. George Serghides, long-serving President of Family Court to the position of Cypriot judge at the European Court of Human Rights has been the proverbial exception that proves the rule.

10.4.2   The Law Office of the Republic: The Attorney-General as an Oracle of the Law?

The omnipresent office of the Attorney General of the Republic (Γενικός Εισαγγελέας) is probably one of the most important legacies of the British colonial era to the modern legal system.

During colonial rule, the Attorney General acted as both the colonial government's legal counsel and the head of colonial lawyers (for a certain period, even at the head of colonial judges). Upon independence, the Constitution established the Attorney General as the first among the 'independent officers' of the Republic. The Constitution consecrated the Attorney General's role as both 'the legal adviser of the Republic and of the President and of the Vice President of the Republic and of the Council of Ministers and of the Ministers'[109] and the officer vested with full prosecutorial powers.[110] The Attorney-General is also legally regarded as the first lawyer (advocate) of Cyprus: apart from being the Honorary President of the Cyprus Bar Association,[111] he also presides over the Disciplinary Council for advocates,[112] the Advocates Pension Fund[113] and the Law Council.[114]

Perhaps more importantly, the Attorney General is the head of the Law Office, or Legal Service (Νομική Υπηρεσία), of the Republic. The Attorney General is the ostensible head of the Service and he claims political responsibility—and credit—for its actions. It goes without saying that it is impossible to monitor in full all activities, especially as the numbers of Law Office staff have increased, and their responsibilities swollen, following Cyprus accession to the European Union.

The office of the Attorney General also acts as legal counsel to the House of Representatives, advises the Foreign Ministry, organises the participation of Cyprus in EU law-making and the implementation of EU law in Cyprus, and represents Cyprus before European and international Courts. The existence of organised legal support and the small sizes involved have not allowed for the creation of dedicated, independent legal advisors for government offices or the legislature. In fact, unlike Greece for example, members of the Law Office are not seconded to ministries. Such a system may lead to complaints about the level of attention being paid by the Law Office to some departments but it also means that the Law Office exercises almost total control over the legal advice dispensed in the public sector, including all legislative initiatives. As a result, the opinion of the Law Office carries great

---

[109]The Constitution of the Republic of Cyprus, art 113(1).

[110]The Constitution of the Republic of Cyprus, art 113(2).

[111]Advocates Law, arts 23(1) and 23(4).

[112]Ibid., art 16(2).

[113]Ibid., art 26, which authorises the Council of the Cyprus Bar Association to issue Regulations, approved by the Council of Ministers, on the creation and operation of the Advocates' Pension Fund. Issued in 1966, the Regulations name the Attorney-General as president of the Fund's Board of Directors.

[114]Ibid., art 3(1).

weight—hence the allusion of acting as an 'oracle of the Law', in a more literal sense than John Dawson's judges.[115] The Law Office may hold a sort of 'supra-constitutional' status, being capable of controlling all three other powers to a lesser or greater extent.

Having said all this, the effective power of the Law Office and its staff in the governance and legal politics of Cyprus may yet be underestimated.[116] Law Office personnel has often been behind power-play initiatives eventually attributed to other actors. The needs of participating in EU governance and implementing European legislation have further increased the power of the Office and its allies and allowed senior staff from the Law Office to be elevated to EU judicial positions.

10.4.3 Notables and the Bar at Large

The third pillar of the establishment consists of the *notables*—the upper echelons and 'old names' of the legal profession. It is not easy to delimit the boundaries of this group, but at the core we may locate the name partners in law firms that can trace their roots to the 'barristers' of the late colonial period or the founding partners of large law firms established in the first two or three decades after independence. The periphery of this group includes senior partners in—and some of the more successful alumni of—the large law firms, primarily in Nicosia and Limassol, and at least some of the officeholders in the Cyprus Bar Association.[117] The *notables* can draw on long experience, big reputations, strong client bases and kinship ties.

The role of the *notables* is, ironically, strengthened by the way in which the legal profession is organised in Cyprus. Even though the distinction between solicitor and barrister never formally existed in Cyprus, the colonial Bar—and its descendants—invested on the barrister tradition. After independence, the most successful entrants to the legal profession adopted a similar attitude, despite their continental university education; their professed pride in what they learned did not prevent them from sending their children to study law in England—effectively raising barriers against and opposing the next generation of upstarts in a manner not unlike what they had themselves been through in their early years in the profession.

---

[115]Dawson (1986). Dawson himself takes his reference from Blackstone (1769) ('depositories of the law, the living oracles').

[116]The only study of the Law Office has been on criminal justice. See Kyprianou (2009). Dr. Kyprianou herself subsequently joined the Attorney General's office.

[117]In the present government, the President of the Republic (a 1960s Athens graduate with a successful law firm in Limassol) could be regarded as belonging to the first group. Two of the ten ministers have their own law firms, having spent their first years in practice working for a major law firm founded soon upon independence. Another minister with a law degree is a former senior civil servant and the wife of an advocate whose family practice goes back to the late 1920s. The other lawyer in the cabinet went quickly into politics and business activities instead of legal practice. All five hold basic law degrees from Greece.

Reclaiming the barrister tradition was a reason why the definition of what constitutes 'practicing as an advocate' only expanded in the 1980s, following the explosion of corporate and fiduciary services, to include legal consulting activities. The structure that had developed by that time allowed the maintenance and con-solidation of the establishment law firms, several of which have expanded into some of the biggest in the Eastern Mediterranean, with branches in Greece and the former Soviet Union. Large numbers of advocates, working in the big law firms, seldom if ever appear in Court. At the same time, just like lawyers employed in government, in-house counsel of businesses are excluded from Bar membership—leading to a stronger need to recourse to outside legal support, usually to the *notables* and other large law firms.

10.5   Conclusions

As my title suggests, Cyprus is a *biggish* small state—not just in terms of medals collected by athletes in the ESSG. The numbers of Cypriot lawyers and Cypriot cases (both standard cases and landmark cases) probably dwarf those in most of the really small jurisdictions—but of course they are nothing compared to bigger states, including Greece (and Turkey).

The relative small numbers of legal professionals do matter for the shape of the law, with regard both to domestic governance and to how the outside world is approached. Being a small country with immediate access to a ready (and success-fully evolving) legal system has meant that it would have been difficult, and possibly inefficient to create a fully independent Cyprus law. Being a small country has led to a strong role for the legal profession as a whole in the country—but also to increased power for relatively small (elite) groups within the legal profession.

Both these factors largely explain why Cyprus is today a mixed jurisdiction where the English legal tradition still holds considerable sway. In fact, it is probably still true that to actually learn and successfully practice Cyprus law, one has to initially approach it in common law terms. We already took note of how, even though the barrister/solicitor distinction was never formally adopted in Cyprus, it is still indirectly alluded to in elite genealogies and lawyerly folklore. Moreover, even legal fields inspired by the Continental legal tradition are *mutating* through the use of common-law procedure and the common-law mentality prevalent among the judiciary and most of the legal elite.

At the same time, one must also account for the fact that mutation patterns are increasingly noticeable even in certain common-law fields, including civil practice. Ready access to Cypriot cases, with immediate authority and written in the native Greek language is certainly a reason: regardless of how the common law doctrines of what constitutes persuasive and what binding authority are played out in Cyprus, and even though many local decisions do not show the intellectual prowess of equivalent English case law, in the very least they can provide an indication of how Cypriot judiciary approaches matters—and, over time, they can provide a sort of *jurisprudence constante* that must be followed. Another reason has to do with Cyprus not having followed, for better or worse, many of the law reforms implemented in England over the past few decades. But one must not underestimate social factors: to begin with, respect for English law is due mostly to pragmatic considerations and power games and does not necessarily translate to Anglophile sentiments for much of the legal profession, let alone the general population. The gatekeepers themselves are over time shifting the grounds for their legitimacy—from advanced knowledge of the common law to their contribution to the uniqueness and intricacies of the native legal system. The legal profession itself is pushing towards mutation and away from common law ideal types. In form and in substance, the Cyprus bar is a massive, unitary body of *advocates* and its professional bodies are more about lawyer politics than the pomp and tradition of Inns and Law Societies. As to the judiciary, for all its common law consciousness (and roots), the system of promotions and transfers—with the Supreme Court justices acting as the Supreme Judicial Council—has established a hierarchical, even bureaucratic judicial system made even more so by the strictest adherence to seniority for appointment to high judicial office. The ubiquitous role of the Supreme Court may have granted great power to that institution but it does not necessarily help it fully achieve the mission of a common-law supreme Court.

Having said all this, one must guard against underestimating the importance of policy, politics, identity and personal ambition in fostering law reform. One might assume that, for example, a lightly populated Commonwealth offshore jurisdiction would be less likely to deviate from the common law than a sizable small state such as Cyprus—with the exception probably of subjects and rules important to that offshore jurisdiction. But the opposite may in fact hold true: the smaller the numbers of the stakeholders that have an effective say, the more homogeneous they are likely to be—thus, the easier it should be for a person or a group in power to alter the status quo.

The case of Cyprus shows that this is partly the case. On the one hand, most law reforms have originated with the gatekeepers identified in the previous section. But there have also been several noticeable instances, even waves, of reform initiated by the legal profession at large. On the whole, it may be that they key word for reform in Cyprus is *consensus*—either broad consensus where a reform is broadly discussed and accepted or passive consensus where reforms are tolerated by the affected constituencies.

The political power yielded by lawyers as a whole has certainly had repercussions for the political development of Cyprus. It has also allowed lawyers or groups within the legal elite to often control, or hinder, law reform. But it has also helped shape a unique Cyprus identity, respectful of but also distinct from the Greek Cypriots' ethnic identity: law in Cyprus is not the expression of the people at large, but it serves as an illustration of how Cypriots have, through their history, attempted and often succeeded to overcome adversities deriving from their island's strategic location and our own mistakes and to even turn them into strategic advantage. Cyprus law in that regard, stands for our own combination of tradition and innovation, insularity and extroversion, lofty aspirations and hard-nosed pragmatism.

References

Artemis P (2004) Prerogative writs: Principles and Cases. Nicosia (in Greek)
Baldersheim H, Keating M (eds) (2015) Small states in the modern world. Edward Elgar, London
Blackstone W (1769) Commentaries on the law of England. Clarendon Press, Oxford
Cassin Ritaine E et al (eds) (2010) Comparative law and hybrid legal traditions. Schulthess, Zurich
Cairns JW (2014) Watson, Walton and the history of legal transplants. Georgia J Int Comp Law
    41:637–696
Castellucci I (2012) Legal hybridity in Hong Kong and Macau. McGill Law J 57:665–720
Dawson J (1986) The oracles of the law. William S Hein & Co, Getzville, New York
European Justice (2016) Legal Professions. Available at https://e-justice.europa.eu/content_legal_
    professions-29-en.do. Accessed 3 Aug 2016
Fedtke J (2006) Legal transplants. In: Smits J (ed) Elgar encyclopedia of comparative law. Edward
    Elgar, London, pp 434–437
Hadjihambis DH (2010) The supreme court of Cyprus and its judges since its establishment in
    1883. Supreme Court of Cyprus, Nicosia
Hatzimihail N (2013) Cyprus as a mixed legal system. J Civil Law Stud 6:37–96
Hatzimihail N (ed) (2017) Modern aspects of Civil Law: Family Law in Greece and Cyprus.
    Nomiki Bibliothiki, Athens (in Greek)
Hatzimihail N (2015) Reconstructing mixity: sources of law and legal method in Cyprus. In:
    Palmer V et al (eds) Mixed legal systems, East and West. Ashgate Publishing, London, pp
    75–99
Hellenic Resources Network (2016) Logos news from Cyprus. Available at http://www.hri.org/
    news/cyprus/logosg/1997/97-09-29.logosg.html. Accessed 23 Aug 2016
Hill G (2010) A history of Cyprus, vols 1–4. Cambridge University Press, London
Kennedy D (1997) A critique of adjudication. Harvard University Press, Massachusetts
Kombos C (2015) The doctrine of necessity in constitutional law. Sakkoulas Publishing, Athens
Kyprianou D (2009) The role of the Cyprus attorney general's office in prosecutions: rhetoric,
    ideology and practice. Springer, Berlin
Mallinson W (2008) Cyprus: a modern history. I B Tauris & Company Ltd, New York
Malliotis A (2010) Contracts. In: Campbell D (ed) Introduction to Cyprus Law, 3rd edn. Yorkhill
    Law Publishing, Salzburg
Ministry of Foreign Affairs of the Republic of Cyprus (2012) Cyprus at a glance. Available at
    http://www.mfa.gov.cy/mfa/mfa2006.nsf/glance_en/glance_en?OpenDocument.   Accessed
    19 Aug 2016
Nicolaou E (1998) Recent developments in family law in Cyprus. In: Bainham A (ed) International
    survey of family law 1006. Martinus Nijhoff, Leiden
Örücü E (2008) What is a mixed legal system: exclusion or expansion? Electronic J Comp Law 12
    (1) Available at http://www.ejcl.org/121/art121-15.pdf. Accessed 20 Sept 2016
Palmer V (2008) Two rival theories of mixed legal systems. Electronic J Comp Law 12(1).
    Available at http://www.ejcl.org/121/art121-16.pdf. Accessed 20 Sept 2016
Palmer V (ed) (2012) Mixed jurisdictions worldwide: the third legal family, 2nd edn. Cambridge
    University Press, Cambridge
Papacharalambous C (2015) Cypriot criminal law: general part, vol I. Nomiki Bibliothiki, Athens
    (in Greek)
Papasavvas SS (1998) La justice constitutionnelle à Chypre. Economica, Paris
Polyviou PG (2015) The case of Ibrahim, the doctrine of necessity and the Republic of Cyprus.
    MAM, Nicosia
Serghides G (1988) Internal and external conflict of laws in regard to family relations in Cyprus.
    GAS Publications, Nicosia
Serghides G (2010) Reflections on some aspects of the family law of the Turkish community in
    Cyprus. GAS Publications, Nicosia
Solsten E (ed) (1991) Cyprus: a country study. GPO for the Library of Congress, Washington

Symeonides SS (2003) The mixed legal system of the Republic of Cyprus. Tulane Law Rev 78:441–455

Teubner G (1998) Legal irritants: good faith in British law or how unifying law ends up in new divergences. Modern Law Rev 61:11–32

The World Bank (2016) Small states. Available at http://www.worldbank.org/en/country/smallstates. Accessed 19 Aug 2016

Theofanous A (2016) Cyprus: from an economic miracle to a systemic collapse and its aftermath. In: Briguglio L (ed) Small states and the European Union: economic perspectives. Routledge, London, pp 28–49

Thorhallsson B (2000) The role of small states in the European Union. Routledge, London

Thorhallsson B (ed) (2004) Iceland and European integration: on the edge. Routledge, London

Thorhallsson B, Wivel A (2006) Small states in the European Union: what do we know and what we would like to know? Camb Rev Int Aff 19:651–668

Trusted HH (ed) (1959) Statute laws of Cyprus, vol 1. C.F. Roworth, United Kingdom

Watson A (1993) Legal transplants: an approach to comparative law, 2nd edn. University of Georgia Press, Georgia

Chapter 11
Choices for the South Pacific Region's Bar Associations and Law Societies?

Nilesh N. Bilimoria

11.1 Introduction

A robust model for the regulation of the legal profession is a necessary condition for upholding the rule of law and maintaining public confidence in the legal and justice systems in any state. And in the South Pacific, a large geographical region peppered with numerous small states, the need for an effective and robust regulatory model for processing and investigating complaints and disciplining legal practitioners cannot be overstated.

This chapter considers the particular problems faced by the legal profession across the small states of the South Pacific and the various regulatory models that may be adopted to deal with ethical and other breaches of professional standards. It commences by providing a snapshot of the concerns expressed about the experiences of, and demands made on, new law graduates engaged in legal practice in the South Pacific region. It extends to providing testimonies of young law graduates as they enter the profession (Sect. 11.2). This is followed by an examination of the available regulatory models for processing, investigating and disciplining legal practitioners in the South Pacific region (Sect. 11.3). Particular attention is drawn to Fiji's disciplinary models pre 2009 and post 2009, which form a basis for other member countries of SPLA (South Pacific Lawyers' Association) to revisit its existing model in setting high standards of professional ethics when delivering legal services to members of the public. Recent developments initiated through SPLA include the formulation of model legal professional conduct rules ('South Pacific Model Conduct Rules'). The 2014 Pacific Legal Profession Survey is also

N.N. Bilimoria (✉)
School of Law, The University of the South Pacific, Suva, Fiji

Faculty of Law, Queensland University of Technology, Brisbane, Australia
e-mail: nileshnirvaan.bilimoria@hdr.qut.edu.au

© Springer International Publishing AG 2017
P. Butler, C. Morris (eds.), *Small States in a Legal World*,
The World of Small States 1, DOI 10.1007/978-3-319-39366-7_11

discussed briefly as an evolving initiative (Sect. 11.4). In short, the South Pacific Model Conduct Rules are intended to set minimum standards for the professional conduct of lawyers in the region and it is further intended to provide a streamlined set of parameters for effective regulation of the profession by member countries. Whether these initiatives will provide the region's Bar associations and law societies concerted direction and comfort to resurrecting community confidence in the legal profession in the region, is yet to materialise and be tested. Not all member countries' current status of the legal profession regulation is discussed, which forms a limitation in this study. A further limitation takes the form of the 12 *newsSPLAsh* issues published by SPLA to date which have been reviewed for this study. Empirical and qualitative research was not conducted at this stage.

In 2011,[1] under the umbrella of the SPLA, as the peak legal professional institution for the region, member law societies and bar associations came together to rewrite the rulebook for raising the conduct bar for the legal profession in the region.[2] At this point, it would be an oversight to fail to acknowledge the formulation of SPLA, in establishing a central platform in the South Pacific region for a well-administered legal professional space for legal practitioners. In 2007, conversations were held by participants comprising of 21 legal practitioners from 11 South Pacific jurisdictions (Cook Islands, Kiribati, Vanuatu, Tuvalu, Niue, Papua New Guinea, Samoa, Fiji, Tonga, Solomon Islands, Noumea, French Polynesia, Timor-Leste and Norfolk Island) with representatives from Law Council of Australia, LAWASIA, New Zealand Law Society, AusAID, and the Queensland Bar Association during its first South Pacific Roundtable in Brisbane, Australia.[3] It was through this forum and subsequent meetings that participants raised common concerns in relation to the status and effectiveness of their own legal professional institutions and the challenges faced by legal practitioners in delivering legal services in light of the population to practitioner ratio in the South Pacific region.[4] The ineffective regulatory framework for disciplining legal practitioners for breaching rules of professional conduct and practice, the need for a uniform guiding framework for professional conduct and structured ongoing continuing legal education were some concerns to name a few.[5]

These concerns, as well as the goals of resurrecting public confidence and encouraging members of the legal profession to adhere to high ethical standards in the legal profession in the South Pacific region, led to the official launch of SPLA at the Commonwealth Law Ministers' meeting in Australia on 10 July 2011.[6]

---

[1] South Pacific Lawyers Association (2011a), p. 12.

[2] Ibid, p. 13. See also Ibid, p. 2; South Pacific Lawyers Association (2011c), p. 5.

[3] See South Pacific Lawyers Association (2011a), pp. 12 & 14–15. See also South Pacific Lawyers Association (2011e), p. 2: The current membership was extended from 11 to 15 member countries.

[4] See South Pacific Lawyers Association (2011a), pp. 14–16.

[5] South Pacific Lawyers Association (2011b), pp. 8–10.

[6] South Pacific Lawyers Association (2011a), p. 13.

11.2 Lawyers in the Pacific, Reactions and Testimonies

The University of the South Pacific (USP) is uniquely positioned in the South Pacific region as the premier institution for higher learning and teaching and research. It offers a number of programs across a variety of disciplines to students from member countries of USP.[7] 'Over the past 15 or so years, hundreds of students have graduated from the University of the South Pacific (USP) with a Bachelor of Laws Degree (LLB) followed by a Professional Diploma in Legal Practice (PDLP). It is estimated that more 80 per cent of the legal profession in the South Pacific are graduates of USP'.[8] These students in turn, return to their respective home countries and move on to practice as a legal practitioner in private practice or otherwise.

In 2013, the Chief Justice of the Solomon Islands, the honourable Sir Albert Palmer, put forth his view that 'young law graduates are going into private practice without proper training and mentoring from more senior and experienced lawyers'.[9] Ross Ray, past Chairperson for SPLA, replied by commenting, 'in his experience, lawyers in the South Pacific are all very well-educated and capable people. . .[but] while lawyers receive strong undergraduate education there is no facility for further professional training'.[10]

Sir Albert's concerns may very well be valid. Sadly, the demands placed on legal practitioners in private practice across the 12 nations that USP services are overwhelming. Ray highlighted the statistics pertaining to the 'ratio of lawyers in private practice to population' in the South Pacific region, for example in 'Fiji 1:4000, Kiribati 1:12,000, Papua New Guinea 1:10,500, and Solomon Islands 1:13,500'.[11] These ratios, he argues, reflect the demands that are placed on young law graduates to deliver legal services to members of the public. During the Pacific Young Lawyers Forum on 10 July 2011 held in Brisbane, one of the participants, Puloka from Tonga reported that:

> Working as a young lawyer in the Pacific has its challenges. The pressure to maintain ethical standards and model conduct is not only limited to working hours, but also extends to the lawyer's personal life. There is a shortage of lawyers in Tonga and a lack of mentoring from senior lawyers in the Pacific.[12]

According to Cartledge, 'USP graduates are often thrust into positions of authority and responsibility beyond the level of their "training" without the support

---

[7]More information available at http://www.usp.ac.fj/. Accessed 10 July 2016. See also, http://www.usp.ac.fj/fileadmin/scripts/HandbookAndCalendar/HandbookAndCalendar_2016_en.pdf for their 2016 Handbook and Calendar. Accessed 10 July 2016, p. 22.

[8]South Pacific Lawyers Association (2011d), p. 6.

[9]Palmer, in South Pacific Lawyers Association (2013), p. 3.

[10]Ray, in ibid.

[11]Ray, in South Pacific Lawyers Association (2011e), p. 5. See also South Pacific Lawyers Association (2013), pp. 3–4 & Appendix I for a breakdown of practitioner numbers across the South Pacific region.

[12]See South Pacific Lawyers Association (2011b), p. 6.

of an experienced mentor. They are literally thrown in at the deep end'.[13] Puloka confirms this, saying, 'there is a great amount of work pressure on young lawyers who have been thrown into the deep end of their legal career. . ..climbing the career ladder has so far been rather quick. Due to the lack of lawyers in the Pacific, young lawyers are forced to fill in the gaps of senior posts in the legal fraternity'.[14]

Cartledge further explains, with all respect directed towards Sir Albert Palmer and other critics:

> Few beyond the South Pacific appreciate the effort, personal sacrifice and determination it takes for most Pacific Island students to pursue an English common law-based law degree which is taught exclusively in English, as for many, English is a second, third or fourth language. Critics may also fail to appreciate the complexities and unique challenges a regional university faces in teaching law. Apart from the difficulties posed by language, many of the subjects must be taught in a generic form, as it is neither practical nor appropriate to impose on students an obligation to acquire a comprehensive appreciation of the laws of all or any particular" jurisdiction of USP's member countries.[15]

Kuautonga, from the Solomon Islands, and another participant at the Pacific Young Lawyers Forum reported that:

> Being a young woman lawyer in the Solomon Islands is a challenge. When you are a student your mind concentrates on one to four assignments in two to three months. Working with the Public Solicitor's Office, one has to deal with whatever matter that is thrust one's way. On a personal level, there is a custom called the wantok system that is very common in Solomon Islands, whereby the position as a public servant is used to serve the benefit of relatives, family members and friends. It is a form of corruption that has been imprinted in the minds of the general public. As long as one is a public servant, you have to serve your wantoks despite the fact that you have certain codes of conduct to abide by. So it's a conflict between performing to the expected standard and using the Office for the best interest of wantoks. . .[16]

Other young lawyers from the region also shared their testimonies which are noteworthy as follows:

> [In] Tuvalu, while no restrictions are placed on practice by female lawyers, cultural barriers exist in that mediation and arbitration are traditionally not seen as women's roles. The Tuvaluan Constitution permits discrimination on the basis of gender and some inequalities are present in relation to family law and property law matters.[17]

> Being in a small, relatively new jurisdiction, you get to help in developing the jurisprudence when you are involved in complex legal cases which go before the courts.[18]

> Kiribati law and the legal profession are still developing and it is a pleasure to take part in the development of these things...Due to the relatively small number of lawyers in Kiribati, the law profession in Kiribati is very close...Although the six months Professional

---

[13]Cartledge, in South Pacific Lawyers Association (2011d), p. 9.

[14]See South Pacific Lawyers Association (2011b), pp. 6–7.

[15]Cartledge, in South Pacific Lawyers Association (2011d), p. 6.

[16]See South Pacific Lawyers Association (2011b), p. 7.

[17]Ibid, p. 8.

[18]See South Pacific Lawyers Association (2011a), p. 10.

Diploma in Legal Practice helped a lot, the first few months working as a lawyer was a steep learning process.[19]

The purpose of providing this snapshot of the concerns of the established legal fraternity and the experiences of young law graduates entering the profession is twofold. Firstly, it demonstrates that despite high expectations from member countries that their law students are offered content and curriculum pertaining to their particular jurisdiction, USP does and continues to deliver a quality LLB programme and post-graduate pre-admission training with coverage across the Pacific region. Secondly, it can be difficult to gauge how fresh graduates fare as they enter the legal profession: their testimonies are valid indicators that fresh graduates have a legitimate expectation that senior practitioners and experienced practitioners will play a gatekeeper role for them, so that they can effectively acquire practical skills and maintain high ethical standards in the profession.

11.3 Models of Regulation in the South Pacific

As noted, the problems inherent in the Pacific's small jurisdictions: a limited number of (senior) practitioners, conflicting cultural and professional expectations, increased potential for conflicts of interest, pressures to take on significant cases at an early point in one's career, and the problem of providing a suitable legal education and continuing career training across jurisdictions mean that practitioners can fall into ethical breaches and/or fall foul of professional standards. These need to be remedied in a way that upholds public confidence in the legal profession as well as providing appropriate discipline to the practitioner.

The discussion that follows on the types of regulatory models available for the legal profession in the South Pacific region draws on the work of Tauvasa Chou-Lee, as featured in Issue 7 of *newSPLAsh* published by SPLA.[20] Chou-Lee's research results in a choice of three models being put forward. These are: (1) fully-independent regulation, (2) self-regulation, and (3) co-regulation. He describes each model as follows:

> [F]irstly a fully independent regulation model is when a legal profession is regulated by a statutory authority that is independent of the legal profession; secondly a self-regulation model is when a legal profession is regulated by a professional association, such as a Law Society; and thirdly, a co-regulatory model is when some regulatory functions are performed by an independent regulator, while other regulatory functions are performed by a professional association.[21]

In analysing these models, he finds that there are several key factors which affect the effective functioning of the professional conduct models in the South Pacific

---

[19]See South Pacific Lawyers Association (2011b), p. 7.

[20]Chou-Lee, in South Pacific Lawyers Association (2013), p. 7.

[21]Chou-Lee, in ibid., p. 13.

region. These, he claims, are, 'costs in regulating the legal profession, the complexity of their framework and operations, their levels of responsibilities and accountabilities, and the speed at which complaints from the public are resolved'.[22] Primarily, most Bar associations and law societies in the South Pacific region are slanted towards the self-regulatory model, including Fiji. However, since 2009, Fiji has revisited its regulatory model and shifted from self-regulatory model to independent regulator model, which is discussed later.

According to Chou-Lee, 'self-regulation usually works with the voluntary support of senior legal practitioners through their time and energy'.[23] In the SPLA's Needs Evaluation Survey and subsequent Needs Evaluation Report released in 2011,[24] its findings on the current administration of South Pacific region's Bar association and law societies confirmed that many of the professional associations were sharing resources, such as office space with another private or government department for its regular operation, functioning and administration. It recommended that additional resources in the form of funding from government and or other agencies would significantly improve the effective operation of bar associations and law societies in the region.[25]

A snapshot of its findings pertaining to the existing infrastructure in the region is noteworthy. For example:

> The Vanuatu Law Society (VLS) sees the establishment of a permanent office as being vital for the proper coordination of its affairs as well as timely responses to issues. Currently, the President and one or two other practitioners are endeavouring to be the driving force behind the VLS, but they cannot devote more time from their practice.[26]

> The Executive members of the Solomon Islands Bar Association (SIBA) are all volunteers with limited time devoted to SIBA activities. A full-time person recruited to administer SIBA would ensure that SIBA matters are given the priority and importance they deserve. An administrator would enable SIBA to provide services for members such as CLE. It is difficult for Executive members to devote the time needed to arrange such sessions when they also practice.[27]

Chou-Lee asserts that 'a professional association as the sole regulator can have limited resources to carry out regulatory functions, including the availability of practitioners who are prepared to prosecute colleagues'.[28] It is only human nature that legal practitioners can feel uncomfortable prosecuting their colleagues and thus 'declare their conflicts of interest and avoid dealing with colleagues who are the subject of complaint or complaints'.[29] As reflected in the Needs Evaluation Report,

---

[22]Chou-Lee, in South Pacific Lawyers Association (2013), p. 13.

[23]Chou-Lee, in ibid., p. 14.

[24]South Pacific Lawyers Association (2011e), pp. 6 & 6–20.

[25]Ibid.

[26]Ibid, p. 18.

[27]Ibid, p. 19.

[28]Chou-Lee, in South Pacific Lawyers Association (2013), pp. 13–14.

[29]Chou-Lee, in ibid., p. 14.

Kiribati for example, has approximately 8 legal practitioners in private practice whilst 25 are in government.[30] In Samoa, approximately 40 practitioners are in private practice whilst 52 are in government.[31] Sadly, this leaves the majority of the complainants (members of the public) with no further avenue to seek redress, unless they are financially secure to escalate their complaints to another level, if this is even available in their jurisdiction. This in turn, diminishes the confidence of members of the public in the legal system and its practitioners as the institution which is legitimately responsible for handling complaints and disciplining legal practitioners has failed to entertain their concerns properly. This brings us to consider the plight of South Pacific region's Bar association and law societies complaints and discipline handling processes under the self-regulatory model.

The Needs Evaluation Report revealed that lack of funding and thinly structured statutory provisions for the regulation of the profession were primarily key factors contributing to Bar associations and law societies for not having in place a structured system to effectively deal with complaints and implicate legal practitioners for breaching provisions of the legal practitioner's legislation in their respective jurisdictions.[32]

A snapshot of its findings pertaining to the statutory framework for handling complaints and discipline across the South Pacific is worth presenting. For example, Nauru, having no private practitioners and approximately 7 practitioners in government, does not have a system in place for complaints and discipline handling.[33] Similarly, for Niue, government lawyers are brought before a complaints handling Ombudsman, but not private practitioners as none exists.[34] It would be safe to state, and at the same time premature to further state that for Nauru and Niue, the demographic information shows its very low membership of the legal profession which does not warrant for such a system to be established.

In the Solomon Islands apart from the only reported case of a legal practitioner prosecuted under the Penal Code of Solomon Islands (note the Legal Practitioners Act),[35] 'no practitioner has yet been sanctioned as a result of a Disciplinary Panel hearing'.[36] The contributing factors have been due to the 'lack of commitment by disciplinary committee members, lack of procedural rules as to how a disciplinary hearing should be conducted, and conflicts on the part of practitioners sitting on the Disciplinary Panel'.[37]

---

[30]See South Pacific Lawyers Association (2011e), p. 14.

[31]Ibid.

[32]See South Pacific Lawyers Association (2013), p. 14. See also South Pacific Lawyers Association (2011e), pp. 25–31.

[33]See South Pacific Lawyers Association (2011e), pp. 25–26.

[34]Ibid.

[35]*Regina v Ashley* [2011] SBHC 169. Available at http://www.paclii.org/sb/cases/SBHC/2011/169.html. Accessed 10 July 2016.

[36]See South Pacific Lawyers Association (2011e), p. 27.

[37]Ibid.

In Samoa, a recent decision of the Western Samoa Supreme Court is noteworthy in the context of the composition of a disciplinary committee and the application of procedural rules in conducting disciplinary proceedings. In the leading case authority of *Tuala Auimatagi Ponifasio v Council of Samoa Law Society*,[38] the Supreme Court awarded the appellate the sum of WST$48,000, quashing the respondent's decision to suspend the appellate from practice on 12 June 2011. In short, the two issues before the Supreme Court were: (1) whether the composition of the disciplinary tribunal (Council of Samoa Law Society) was independent and (2) whether the respondent was accorded procedural fairness during the respondent's disciplinary hearing before the disciplinary tribunal and a right to fair trial in line with Article 9 of the Constitution of Samoa.[39]

11.3.1 Models of Regulation: The Case of Fiji

This brief overview of Pacific models of legal professional regulation brings us to discuss the fully independent regulatory model, commonly taking the form of a semi-autonomous body.

As Chou-Lee points out, such a body 'focuses more on specific types of regulation of the legal profession and would perceptively be more apolitical and impartial than a self-regulated one when it comes to responsibility and accountability'.[40] To validate the nature and scope of such a model, the shift in the disciplinary model in Fiji since 2009 to this form, is noteworthy in many ways compared to other member countries' self-regulatory models. Its features include: (1) published disciplinary rulings together with reasons for establishing a disciplinary count against a legal practitioner by the Independent Legal Services Commission (ILSC) since 2009, and (2) the range of Orders (including setting a tariff for particular types of offending) made against practitioners in Fiji for unsatisfactory professional conduct and professional misconduct. Note that the Fiji Law Society, since its inception till 2009, was responsible for all affairs of the profession pertaining to practitioners' membership, issuance of annual practising certificates, disciplining of practitioners and organising the Society's annual conference. It is still in existence but these functions are now absorbed by the ILSC pursuant to the Legal Practitioners Decree 2009 (LPD).[41] In relation to the LPD,

---

[38][2013] WSSC 1. Available at http://www.paclii.org/ws/cases/WSSC/2013/11.html. Accessed 10 July 2016.

[39]Ibid, paras 6, 11–12. See also Keresoma (2013).

[40]Chou-Lee, in South Pacific Lawyers Association (2013), p. 13.

[41]Available at http://www.paclii.org/fj/promu/promu_dec/lpd2009220. Accessed 10 July 2016. See, in particular, s 184(1) which repeals the Legal Practitioners Act 1997 and s 84. See also http://www.southpacificlawyers.org/member/republic-fiji-islands. Accessed 10 July 2016.

Commissioner (formerly Justice) Paul Madigan, maintains in one of his disciplinary rulings that:

> With the well-known history of the Fiji Law Society' inactivity on complaints against practitioners, the Legislature is, by this Decree, merely putting in place an alternative administrative procedure for supervision of the profession and for the resolution of complaints. There is in the Decree no new draconian duties, not penalties for omissions placed on practitioners that would unfairly affect them in retrospect. The Decree applies to conduct of practitioners when complained of or detected be that conduct in the intermediate past or in the present day.[42]

Before proceeding further, by way of background, the LPD establishes the ILSC that takes care of disciplining legal practitioners when found to be in breach of the provisions of the LPD or any other written law such as the Trust Accounts Act 1996, or the High Court Rules 1998.[43] The external regulation of the profession in Fiji takes its cue from the New South Wales framework for regulating lawyers and enforcing professional ethical standards.[44] The statutory provisions and processes for complaints and discipline handling are briefly examined below:

Firstly, any person can lodge a complaint to the Chief Registrar (Chief Registrar of the High Court of Fiji)[45] and where the Attorney-General or the Fiji Law Society receives a complaint, it can be referred to the Chief Registrar.[46] Secondly, upon a finding of unsatisfactory or unprofessional misconduct by the ILSC, the Order of the ILSC is filed with the High Court of Fiji within 14 days, making it an Order of the High Court.[47] Finally, appeal mechanisms pursuant to the Appeal Rules are also available to a legal practitioner to pursue in the appellate Courts.[48]

The data provided by the Chief Registrar of High Court of Fiji and an analysis of this data (below), confirms: (1) the effectiveness of complaints handling, (2) disciplining practitioners, and (3) positive outcomes for members of the public so that they can place their trust and confidence in the legal profession under a fully independent regulatory model (Table 11.1).

---

[42]*Chief Registrar v Adish Kumar Narayan No. 9 of 2013* [Ruling] (25 Sept 2013), p. 5, para 27.

[43]See Legal Practitioners Decree 2009. See also Law Council of Australia (2013), p. 18.

[44]NSW Law Reform Commission (1993), pp. 71–129. For more detail, see NSW Law Reform Commission. Available at http://www.lawreform.justice.nsw.gov.au. Accessed 10 July 2016; Office of the Legal Services Commissioner. Available at http://www.olsc.nsw.gov.au. Accessed 10 July 2016. See also Administrative Decisions Tribunal Act 1997. Available at http://www.legislation.nsw.gov.au/maintop/view/inforce/act+76+1997+cd+0+N. Accessed 10 July 2016; Law Council of Australia (2013).

[45]Legal Practitioners Decree 2009, s 99. See also Law Council of Australia (2013).

[46]Legal Practitioners Decree 2009, s 102. See also Law Council of Australia (2013).

[47]Ibid, s 122(2). See also Law Council of Australia (2013).

[48]*Chief Registrar v Abhay Singh Independent Legal Services Commission No 1 of 2009* [Judgment] (25 Jan 2010). See also *Abhay Kumar Singh v Chief Registrar of the High Court No 3 of 2010 Court of Appeal of Fiji* [Judgment] (22 Sept 2010); *Abhay Kumar Singh v Chief Registrar of the High Court of Fiji No 7 of 2010 Supreme Court of Fiji* [Judgment] (20 Oct 2011); Law Council of Australia (2013).

Table 11.1 Complaints against practitioners: Fiji [**Credit**: The Chief Registrar of the High Court of Fiji – 4/9/15]

Year	Total no. of complaints received	Complaints dealt with mediation	Prosecution	Total dealt with for that year only
2009	411	3	9	12 [399]
2010	332	53	23	76 [256]
2011	300	57	5	62 [238]
2012	307	120	10	130 [177]
2013	333	119	39	158 [175]
2014	271	140	17	157 [114]
2015 (31st Aug)	159	91	3	94 [65]

An analysis of the data is provided as follows:

In 2009, the Chief Registrar of the High Court inherited 300 files containing complaints against legal practitioners in Fiji from the Fiji Law Society. In addition to these complaints, the Chief Registrar received 111 fresh complaints against practitioners, making a total of 411 to be dealt with in 2009. The number of fresh complaints against legal practitioners from 2010 to 2014 is on an average of approximately 300 complaints during each single year. The numbers in closed brackets in the last column mean that, that many number of complaints are still pending for that particular year, and by adding all the brackets together for each of the years, the total number of complaints pending as at 31 August 2015, will add up to 1424.

Mediation is also a feature under this type model, allowing complaints to be resolved informally and every year since 2009 the complaints resolved by mediation have increased. The question can be asked why all complaints or majority of the complaints cannot be resolved during that year of receiving such complaints? The response rests on how evenly the budgetary allocation is spread in light of allocation to other departments, agencies and sectors by the government of Fiji. The overall impression of the disciplinary response to breaches by practitioners in Fiji under the independent regulator model can be seen to be as an improvement in public confidence and effectiveness in dealing with complaints.

Furthermore, the effectiveness of a fully independent regulatory model is illustrated by the types of professional disciplinary cases that have come before the Independent Legal Services Commission. Since 2009, a review of the professional disciplinary cases reveals that routine 'conveyancing transactions and irregularities in professional trust account' to name a few, are rich areas for complaints against practitioners in Fiji.[49] In *Chief Registrar v Vipul Mishra, Mehboob Raza,*

---

[49]See *Chief Registrar v Vipul Mishra, Mehboob Raza, Muhammad Shamsud-Dean Sahu Khan, Sahu Khan & Sahu Khan Independent Legal Services Commission No 2 of 2010* [Judgment] (3 Mar 2011); *Chief Registrar v Divendra Prasad Independent Legal Services Commission No 3 of 2011* [Sentence] (7 Mar 2012). See also Law Council of Australia (2013).

Muhammad Shamsud-Dean Sahu Khan and Sahu Khan & Sahu Khan, the then Commissioner (formerly Justice) John Connors, at paragraph 1 remarked, 'This litany of disaster commenced with what should have been a routine conveyancing transaction. It is difficult to conceive that an innocent member of the community could be treated in the way this complaint was by a brace of senior lawyers'.[50] Commissioner Connors warned:

> In a country such as Fiji where the literacy and understanding is not as high as in developed countries the position held by a legal practitioner is even more special and the responsibilities are even greater. . .It follows from to the authorities that the seniority and notoriety of the 3rd Respondent exacerbates the conduct and does not mitigate it. . .the public must be protected from conduct of the type displayed by the 3rd Respondent. . . .[51]

Likewise, 'failure to respond', disciplinary cases have also surfaced lately, that has led the ILSC to remind practitioners of their duty to maintain high ethical standards in their practice and, equally, to treat or take the regulatory body for practitioners in Fiji seriously. Commissioner (formerly Justice) Paul Madigan, in his ruling in *Chief Registrar v John Rabuku,* in what I describe as three step cautionary reminder to practitioners, warned that:

> [F]ailure to respond to the Registrar (Chief Registrar of the High Court) is not only in direct contravention to the stipulation in section 105 (Registrar may require explanation) of the Legal Practitioners' Decree but it is also showing complete distain and disregard for the authority of the regulatory arm of the profession. Should such practice go unchecked then the profession would become totally unmanageable with the public then being protected and the spirit of the legislation defeated. . .(and later). . .

> The Commission regards non-compliance with the Chief Registrar's requests and demands are very serious failures on the part of the practitioner. If a practitioner cannot regulate his/her own affairs, how can he regulate the affairs of his clients?..(and still later),. . .To defy authority and in doing so to contravene the provisions of Division 3 of the Legal Practitioners' Decree 2009, calls into question the practitioners' suitability to be in practice. In its role of guardian of professional standards, the Commission has no option but to suspend the Respondent's right to practice.[52]

This clearly demonstrates the gravity of non-compliance with the legislative duty stipulated in the LPD. Other grounds for complaints include:

---

[50]See *Chief Registrar v Vipul Mishra, Mehboob Raza, Muhammad Shamsud-Dean Sahu Khan, Sahu Khan & Sahu Khan Independent Legal Services Commission No 2 of 2010* [Judgment] (3 Mar 2011); *Chief Registrar v Divendra Prasad Independent Legal Services Commission No 3 of 2011* [Sentence] (7 Mar 2012). See also Law Council of Australia (2013).

[51]*Chief Registrar v Vipul Mishra, Mehboob Raza, Muhammad Shamsud-Dean Sahu Khan, Sahu Khan & Sahu Khan Independent Legal Services Commission No 2 of 2010* [Judgment on Sentence – 3rd Respondent] (4 May 2011), paras 25, 48–50. See also Law Council of Australia (2013).

[52]*Chief Registrar v John Rabuku Independent Legal Services No 13 of 2013* [Judgment and Sentence] (30 Jul 2013) para 4. See also Law Council of Australia (2013).

- deceptive advertising;[53]
- signing, affixing and witnessing Commissioner for Oaths stamp without a valid practising certificate;[54]
- making appearances in Court without a valid practising certificate;[55]
- providing instructions to appear without a valid practising certificate;[56]
- raising one's voice in Court and showing disrespect to the bench;[57] and
- attacking the reputation of the opposing client.[58]

The range of Orders made against practitioners in Fiji for unsatisfactory professional conduct and professional misconduct under the independent regulatory model and the powers available to the Commissioner are provided for in section 121 of the LPD. A range of the Orders made by the Commissioner from few selected disciplinary cases are set out below:

11.3.1.1 'Failure to Respond'

The penalty that may be imposed by the Commission includes a public reprimand, a fine and suspension from practice ranging from 1 to 3 months. The gravity of offending is treated as professional misconduct. In the most recent failure to respond case, *Chief Registrar v Anand Singh*, the Commission exercised its powers pursuant to section 121 of the LPD to impose penalties, and it was in this ruling that the Commissioner announced that:

> Previous decisions of this Commission have established a 'tariff' for this dereliction of legislative duty (sections 104, 105, 108); that being suspension of practice from one to three months. In [Matter No 14 of 2013,] a one month suspension was ordered for a practitioner who admitted his failing from the very beginning and was most remorseful before the Commission. In [Matter No 13 of 2013] the practitioner displayed a total lack of remorse and offered an excuse to the Commission that was not only unreasonable but which was arrogantly disdainful of his own client. His certificate was suspended for a period of three months.[59]

---

[53]*Chief Registrar v Mohammed Azeem Ud-Dean Sahu Khan Independent Legal Services No 16 of 2013* [Judgment and Sentence] (30 Jul 2013). See also Law Council of Australia (2013).

[54]*Chief Registrar v Kelera Baleisuva Buatoka Independent Legal Services No 20 of 2013* [Judgment] (11 Oct 2013). See also Law Council of Australia (2013).

[55]*Chief Registrar v Siteri Adidreu Cevalawa Independent Legal Services No 6 of 2011* [Extempore Ruling on Sentence] (5 Dec 2011). See also Law Council of Australia (2013).

[56]*Chief Registrar v Adi Kolora Naliva Independent Legal Services No 4 of 2011* [Extempore Judgment] (5 Dec 2011). See also *Chief Registrar v Kini Marawai and Marawai Law Independent Legal Services No 6 of 2012* [Judgment and Sentence] (15 May 2013); See also Law Council of Australia (2013).

[57]See also *Chief Registrar v Alena Koroi Independent Legal Services No 5 of 2011* [Sentence] (14 Mar 2012). See also *Chief Registrar v Amrit Sen Independent Legal Services No 10 of 2013 [Judgment] (6 Nov 2013)*; Law Council of Australia (2013).

[58]See also *Chief Registrar v Savenaca Komaisavai Independent Legal Services No 21 of 2013* [Judgment] (8 Oct 2013). See also Law Council of Australia (2013).

[59]*Chief Registrar v Anand Singh Independent Legal Services No 24 of 2013* [Judgment] (7 Nov 2013), para 33. See also Law Council of Australia (2013).

11.3.1.2 'Practising Without a Valid Practising Certificate'

The pattern of penalty varies as each case turns on its own peculiar circumstances, according to whether the complained-of conduct is treated as professional misconduct or unsatisfactory professional conduct. In *Chief Registrar v Niko Nawaikula and Savenaca Komaisavai*, the Commissioner in publicly reprimanding both respondents, fining the 1st Respondent $2000, and suspending the practising certificate of the 2nd respondent for 3 months, stressed that:

> In a situation where an unlicensed practitioner is wanting to operate under the 'umbrella' of a licensed practitioner almost invariably he or she will be 'freelance' and not in situ in the licensing practitioner's office and he or she will therefore be beyond the influence and control of the licensing practitioner thereby creating the probability of unchecked disorder. The provisions of the LPD (and particularly the provisions relating to licensing) exist to bring order and control over the practice of the profession and the conduct of the individual practitioners and any deviation from this or disregard to the strict provisions will lead to professional anarchy.[60]

Making appearances without a valid practising certificate before the Bench, is in effect a strict liability offence, and the provisions of the LPD extend to making such offending a criminal offence, with a maximum penalty of $5000 and imprisonment for one year for subsequent offending.[61]

11.3.1.3 'Repeat Offending and Borderline Case of Dishonesty'

Regrettably, these cases allow the Commission to strike off the practitioner from the Roll of Practitioners. In *Chief Registrar v Kini Marawai and Rajendra Chaudhry*, the Commission summarised the position as follows:

> Striking off is a sanction reserved for repeated misconduct or a pattern of misconduct calling into question the practitioner's ability to ever be fit and proper: suspension is to be reserved for conduct that is isolated, caused by illness or unsoundness of mind, conduct that is explicable by its own circumstances and which will not necessarily occur again.[62]

11.3.1.4 'Trust Account Irregularities'

Failing to administer accurate professional trust account records, may be a standalone instance of misconduct or can be compounded by other offending. Depending

---

[60] *Chief Registrar v Niko Nawaikula and Savenaca Komaisavai Independent Legal Services No 9 of 2012* [Judgment and Sentence] (12 Apr 2013), para 22. See also Law Council of Australia (2013).

[61] Legal Practitioners Decree 2009, s 52(2). See also *Chief Registrar v Laisa Lagilevu Independent Legal Services No 1 of 2012* [Decision] (16 Mar 2012), para 3; Law Council of Australia (2013).

[62] *Chief Registrar v Kini Marawai and Rajendra Chaudhry Independent Legal Services No 2 of 2012* [Sentence] (5 Oct 2012), para 21. See also Law Council of Australia (2013).

on the circumstances, the Commission can make a finding of professional misconduct or unsatisfactory professional conduct. In *Chief Registrar v Haroon Ali Shah,*
the Commissioner, in striking the name of the respondent off the Roll, fining and
imposing payment of witness expenses and wasted costs stated:

> Trust account abuse is probably the most serious departure from the professional duties a
> practitioner owes to the public. The very purpose of a solicitor's trust account is to protect a
> client's fund and to ensure that those funds are kept safe and applied to the purpose for
> which they were entrusted to the practitioner. Any departure from those purposes is very
> serious indeed and a defalcating practitioner must be visited with penalties of the severest
> degree. Protection of the public must be the paramount consideration over and above the
> livelihood and freedom of practice of any practitioner.[63]

The data, disciplinary cases and the penalties which are imposed on legal
practitioners under the independent regulator model, provides adequate ammuni-
tion to inform other member countries with SPLA, the effectiveness of choosing
such a model, and setting high standards of professional ethics in the regulation of
the legal profession in their respective jurisdictions.

11.3.2 Models of Regulation: Co-regulation

Turning now to a co-regulatory model, the final type of regulatory model where the
regulatory framework of the profession makes provision for a division of duties
between an independent regulator and the professional association. Regrettably,
none of the other member countries disciplinary model in the South Pacific region
fits this model. Thus, further comment cannot be made in this chapter. However, it
does not preclude us from highlighting key areas for future investigation such as:
(1) funding, (2) the nature and scope of complaints handling and disciplining
practitioners, (3) formulation of independent statutes (independent regulator and
for professional association), and (4) continuing legal education for members of the
profession. How and when, if any member country shifts from its existing model to
this model, is yet to materialise.

11.4   South Pacific Model Conduct Rules and Pacific Legal
Profession Survey

As part of its work on professional conduct regulation, SPLA initiated a regional
framework for the legal professional regulation for its member countries in the
South Pacific region. This initiative was discussed during its inaugural roundtable

---

[63]*Chief Registrar v Haroon Ali Shah Independent Legal Services No 7 of 2012* [Sentence] (22 Jun
2012), para 37.

in 2007[64] which received support for the formulation of the South Pacific Model Conduct Rules. At the outset, the objective of this initiative was to 'develop model rules, legislation and regulations which can be adopted with appropriate debate and modification to harmonise regulation of the legal profession in South Pacific countries'.[65] The first phase of the project concluded towards the end of 2014, which resulted in the production of a draft report, the preliminary findings of which were reported during the annual meeting of Pacific Islands Law Officers' Network (PILON) held in Kiribati, 11–12 November 2014.[66] The efforts displayed by SPLA need to be commended as it is a significant task to analyse existing rules and procedures of all member countries in the region.

Briefly, the preliminary findings listed a number of shortcomings whilst compiling the report, which justifies the delays caused in compiling such a report of such a magnitude. The shortcomings were as follows:

- definitional issues—such as 'professional conduct';
- a lack of nexus between findings of misconduct and the disciplinary consequences;
- a lack of vetting mechanisms to deal with complaints,
- no publication of findings or reasons for findings;
- admission requirements; and
- licensing requirements and costs disclosure requirements.[67]

Subsequent phases will follow as SPLA prepares for further dialogue with PILON and other agencies on legal education and effective regulation of the profession in the South Pacific region.

Likewise, SPLA released its Pacific Legal Profession Survey 2014, which 'incorporated recommendations and questions from the 2011 Needs Evaluation Survey and the Women in Law Survey released in 2014'.[68] Briefly, the 2014 Survey is intended to: 'i) collect data on the variety of work performed in the Pacific; ii) collect demographic data on Pacific legal professionals; iii) collect information on CLE (continuing legal education), complaints and discipline and pro-bono culture; and iv) identify challenges and opportunities facing peak legal professional bodies with a few to developing proposal for funding and support'.[69]

It is further intended that these initiatives will set minimum standards for bar associations and law societies for providing quality legal services to members of the public and strengthening the regulation of the profession in the South Pacific region in the very near future.

---

[64]See South Pacific Lawyers Association (2013), p. 9.

[65]South Pacific Lawyers Association (2014), p. 4.

[66]Ibid, p. 5.

[67]Ibid.

[68]Ibid.

[69]Ibid.

11.5   Conclusion

In my view, this review shows the choices for professional regulation for the South
Pacific region's Bar associations and law societies are sufficient to meet the
concerns surrounding the problems faced by legal practitioners and their clients
in small jurisdictions. The question remains whether bar associations and law
societies are confident to take the leap to improve their professional regulation
models? This however, rests on a number of key factors as highlighted earlier
(infrastructure, resources, training), as each bar association and law society has its
own set of challenges.

Self-regulation, which is the more widely existing model used to date amongst
member law societies and bar associations in the region, offers a divided response
in terms of regulating the profession and disciplining lawyers effectively. Retaining
or shifting from self-regulation rests with the people of the profession who practise
in it, and who are positioned uniquely in the region to strengthening or resurrecting
community confidence in the profession. The case study of the external disciplinary
model in Fiji shows that there is high confidence in setting and maintaining high
standards of professional ethics and practice for practitioners in Fiji. This emerging
model has given renewed impetus to public confidence in the profession. It should
be seriously considered by the remaining professional bodies in other jurisdictions
as the best way forward for professional regulation.

Appendices

Appendix 1: Size of National Law Associations in the South Pacific

	National Law Associations in the South Pacific	Private	Public	Total
1	Cook Is	32	12	44
2	Fiji Law Society	230	296	526
3	Kiribati Law Society	12	29	37
4	Nauru Law Society	0	7	7
5	Niue Lawyers	1	5	6
6	Norfolk Is Bar Association	4	5	9
7	Papua New Guinea Law Society	591	288	879
8	Samoa Law Society	40	52	92
9	Solomon Is Bar Association	42	48	100
10	Tonga Law Society	41	11	52
11	Tuvalu Lawyers	1	10	11
12	Vanuatu Law Society	40	37	77

South Pacific Lawyers Association (2011e), pp. 5, 14

Appendix 2: National Associations in the South Pacific Complaints/Discipline Handling Systems

	National Law Associations in the South Pacific	Complaints/Discipline Handling System in Place?		Body/Institution Disciplining
		Yes	No	
1	Cook Is	X		* CJ High Court of Cook Is – receives/investigates/disciplines * Cook Islands Law Society – no role in receives/investigates/disciplines
2	Fiji Law Society	X	X	* FLS – receives/investigates/disciplines (mid 2009) * LPU under CR High Court – receives/investigates/disciplines
3	Kiribati Law Society	X		* The Council (members of KLS) establishes a Professional Conduct Committee (AG, lawyer, lay person)
4	Nauru Law Society		X	* Rests with the jurisdiction of the Courts
5	Niue Lawyers		X	* 1x Prvt. lawyer – Silent 5x Govt Lawyers <Niue Ombudsman>
6	Norfolk Is Bar Association	X		* NBA – Professional Conduct Board – Supreme Court of Norfolk Is
7	Papua New Guinea Law Society	X		* Lawyers Statutory Committee (AG, member of Council of the Law Society>, 3x lawyers, 2x non lawyers)
8	Samoa Law Society	X		* Council of the Samoa Law Society establishes the Complaints and Investigation Committee (members of the Society) to Lawyers Disciplinary Tribunal (senior lawyer (8 years), lawyer (5), non-lawyer)
9	Solomon Is Bar Association	X		* SIBA – AG – establishes Disciplinary Panel (AG Chair)
10	Tonga Law Society	X		* TLS establishes Disciplinary Committee (AG, CJ, member of the Bench, Secretary of the Society)
11	Tuvalu Lawyers	X		* 1x Prvt. Lawyer (Not Available – Peoples Lawyer Act 2008), 10x Govt. (Office of the AG)
12	Vanuatu Law Society	X		* Law Council (CJ, AG, 1x Lawyer) appoint Disciplinary Committee (judicial officer, 1x lawyer, 3x non-lawyers)

South Pacific Lawyers Association (2011e), p. 25

Appendix 3: Complaints Against Practitioners: Fiji

Year	Total number of complaints received	Dealt with through mediation	Prosecution	Total dealt with	# of LOs # IO
2009	411 (300 files inherited from FLS)	3	9	12 [399]	1 and 1
2010	332	53	23	76 [256]	2 and 1
2011	300	57	5	62 [238]	2 and 1
2012	307	120	10	130 [177]	5 and 3
2013	333	119	39	158 [175]	4 and 2
2014	271	140	17	157 [114]	4 and 3
2015 (31 Aug)	159	91	3	94 [65]	3 and 3
	Pending			1424	

Source: Chief Registrar – Legal Practitioners Unit – 4/9//2015

Appendix 4: Professional Association Membership

	National Law Associations in the South Pacific	Private	Population	Ratio
1	Cook Is	32	11,124	1:348
2	Fiji Law Society	230	888,125	1:3840
3	Kiribati Law Society	12	100,743	1:12,593
4	Nauru Law Society	0	9322	N/A
5	Niue Lawyers	1	1311	1:1311
6	Norfolk Is Bar Association	4	2169	1:542
7	Papua New Guinea Law Society	591	6,187,591	1:10,470
8	Samoa Law Society	40	193,161	1:4829
9	Solomon Is Bar Association	42	571,890	1:13,616
10	Tonga Law Society	41	105,916	1:2583
11	Tuvalu Lawyers	1	10,544	1:10,544
12	Vanuatu Law Society	40	224,564	1:5614

Source: South Pacific Lawyers Association (2011e), pp. 5, 14

Appendix 5: Trust Account Requirements

	National Law Associations in the South Pacific	Are Legal Practitioners Required to Hold Trust Accounts?	Yes	No
1	Cook Is			X
2	Fiji Law Society		X	
3	Kiribati Law Society			X
4	Nauru Law Society			X
5	Niue Lawyers			X
6	Norfolk Is Bar Association			X
7	Papua New Guinea Law Society		X	
8	Samoa Law Society		X	
9	Solomon Is Bar Association			X
10	Tonga Law Society			X
11	Tuvalu Lawyers			X
12	Vanuatu Law Society			X

Source: South Pacific Lawyers Association (2011e), p. 40

References

Keresoma L (2013) Court awards lawyer $48,000 in case against Law Society. In: Talamau Online News. Apia, Samoa Islands. Available at http://www.talamua.com/court-awards-lawyer-48000-in-case-against-law-society. Accessed 10 July 2016

Law Council of Australia (2013) Addressing legal challenges in paradise. Law Council Rev:1–72. Available at http://www.lawcouncil.asn.au/lawcouncil/images/LawCouncilReview/Issue%208%20Web.pdf. Accessed 10 July 2016

NSW Law Reform Commission (1993) Report no 70 – scrutiny of the legal profession: complaints against lawyers. Available at http://www.lawreform.justice.nsw.gov.au/Documents/report_70.pdf. Accessed 10 July 2016

South Pacific Lawyers Association (2011a) newSPLAsh Issue 1 (April–June 2011). Available at http://www.southpacificlawyers.org/files/uploads/newSPLAsh%20Issue%201%20%28web%29.pdf. Accessed 11 July 2016

South Pacific Lawyers Association (2011b) newSPLAsh Issue 2 (July–Sept). Available at http://www.southpacificlawyers.org/files/uploads/newSPLAsh%20Issue%202.pdf. Accessed 10 July 2016

South Pacific Lawyers Association (2011c) newSPLAsh Issue 3 (October–December 2011). Available at: http://www.southpacificlawyers.org/files/uploads/018CLA_newSPLAsh_FA_WEB.pdf. Accessed 10 July 2016

South Pacific Lawyers Association (2011d) newSPLAsh Issue 3 (October–December 2011). Available at http://www.southpacificlawyers.org/files/uploads/018CLA_newSPLAsh_FA_WEB.pdf. Accessed 10 July 2016

South Pacific Lawyers Association (2011e) Needs evaluation survey for south pacific lawyer associations – final report. Available at http://www.southpacificlawyers.org/files/uploads/2011%2010%2018%20-%20Needs%20Evaluation%20Survey%20Final%20Report.pdf. Accessed 10 July 2016

South Pacific Lawyers Association (2013) newSPLAsh Issue 7. Available at http://www. southpacificlawyers.org/files/uploads/newSPLAsh%20Issue%207_FINAL.pdf.   Accessed 10 July 2016

South Pacific Lawyers Association (2014) newSPLAsh Issue 11. Available at http://www. southpacificlawyers.org/files/uploads/newSPLAsh%20Issue%2011_FINAL%20web.pdf. Accessed 10 July 2016

Lightning Source UK Ltd.
Milton Keynes UK
UKHW020153020822
406689UK00004B/617